SECOND EDITIO

I0034170

How to Teach Math to Children

By Joohi Lee

UNIVERSITY OF TEXAS - ARLINGTON

cognella® | ACADEMIC PUBLISHING

Bassim Hamadeh, CEO and Publisher
Kassie Graves, Director of Acquisitions
Jamie Giganti, Senior Managing Editor
Jess Estrella, Senior Graphic Designer
Marissa Applegate, Senior Field Acquisitions Editor
Natalie Lakosil, Licensing Manager
Kaela Martin and Rachel Singer, Associate Editors
Allie Kiekhofer, Interior Designer

Cover image copyright © Depositphotos/leporiniumberto.

Printed in the United States of America

ISBN: 978-1-5165-0347-6 (pbk) / 978-1-5165-0348-3 (br)

cognella® | ACADEMIC PUBLISHING

Table of Contents

School Mathematics for Young Children

At the end of Chapter 1, you should be able to:

- Describe what children need to learn mathematics;
- Identify National Council of Teachers Mathematics (NCTM) content and process standards of mathematics;
- Define NCTM's six principles of teaching mathematics.

1.1 Why Mathematics?

Everyday life requires mathematics skills in many different ways: comparing prices when purchasing, calculating tips, planning for road trips by car (traveling time, distance), and so forth. Mathematics cannot be separated from real life. Mathematics is also an integral part of the human "intellectual heritage" and provides the foundation for how we think and live in current society. Furthermore, most workplaces require a certain level of mathematics or quantitative reasoning skills. These skills are critical in careers associated with high technology or scientific skills (e.g., engineers, scientists, technicians, etc.).

Mathematics is everywhere in everyone's lives regardless of what we do. It is everywhere in every child's real life as well. Children see mathematics everywhere they go, in different shapes, numbers, patterns, and so forth. Children are constantly exposed to mathematics. There is no question that mathematics is an important subject for all children to learn, but what to teach and how to teach it are still debated by educators.

1.2 What to Teach and How to Teach

National, state, and school district guidelines, including standards and district curriculum, help educators to determine what to teach. From a broad perspective, society's requirements related to mathematics also impact decisions about what to teach. A look at the history of mathematics education in the United States shows that focal mathematics content has changed over the years.

In the 1920s, the Progressive movement was prevalent. Progressives believed children would learn arithmetic as they needed to. This was called "incidental learning." During the late 1920s, the Committee of Seven had a strong impact on mathematics education, especially the sequence of what to teach. The Committee of Seven originally consisted of school superintendents and principals. Based on their experience, they introduced the concept of children's mental age and recommended teaching mathematics based on mental age. For example, subtraction facts under ten should be taught to children with a mental age of six years and seven months. In the mid-1930s, Gestalt theory influenced mathematics education by placing more importance on the "understanding" of relationships, structures, patterns, or principles. Based on Gestalt theory, drill is not meaningful if it is not accompanied by understanding. When the Soviet Union launched the first Sputnik satellite into orbit in 1957, other countries turned their attention to technology, especially the United States. As a result, "New Math" was taught during the 1950s and 1960s, focusing on unifying themes to advance the field of technology. Current trends of mathematics education are related to college and career readiness throughout the school years.

This evolution shows that mathematics teaching techniques have changed based on the needs of social entities. However, how to teach mathematics depends on the philosophy of individual teachers, which is influenced by current research findings as well as developmental theories (see Chapter 3 for further discussion). There have been tremendous efforts among educators, researchers, and policymakers to improve the quality of teaching mathematics. Correlations between the quality of teaching and student mathematics achievement have been clearly shown. As a result, many turn their attention to the quality of mathematics teaching. According to the National Council of Teachers of Mathematics (2000), to provide more meaningful mathematics learning experiences for children, teachers of young children should be able to connect mathematics with children's real lives. When children see and become aware of how mathematics works in their real lives, they will be more interested in learning mathematics.

In addition, teachers of young children need to teach important mathematics involving focal points presented by the National Council of Teachers of Mathematics (NCTM), the leading organization of mathematics education in the United States. NCTM is recognized as the most prestigious professional organization in the field of mathematics education and is composed of teachers, educators, professors, and researchers. NCTM first published its *Curriculum and Evaluation Standards for School Mathematics* in 1989. In 2000, these standards were updated and published as *Principles and Standards for School Mathematics*. These updated standards included pre-K school levels for the first time and currently apply to most public school curricula in the United States.

The five content and five process standards for pre-K through grade 12 mathematics education provide mathematics teachers with comprehensive guidelines for what to teach in terms of content and processes along with the expectations of what students should know and be able to do.

The content and process standards are based on grade band (pre-K through grade 3, grades 4 and 5, grades 6 through 8, and grades 9 through 12).

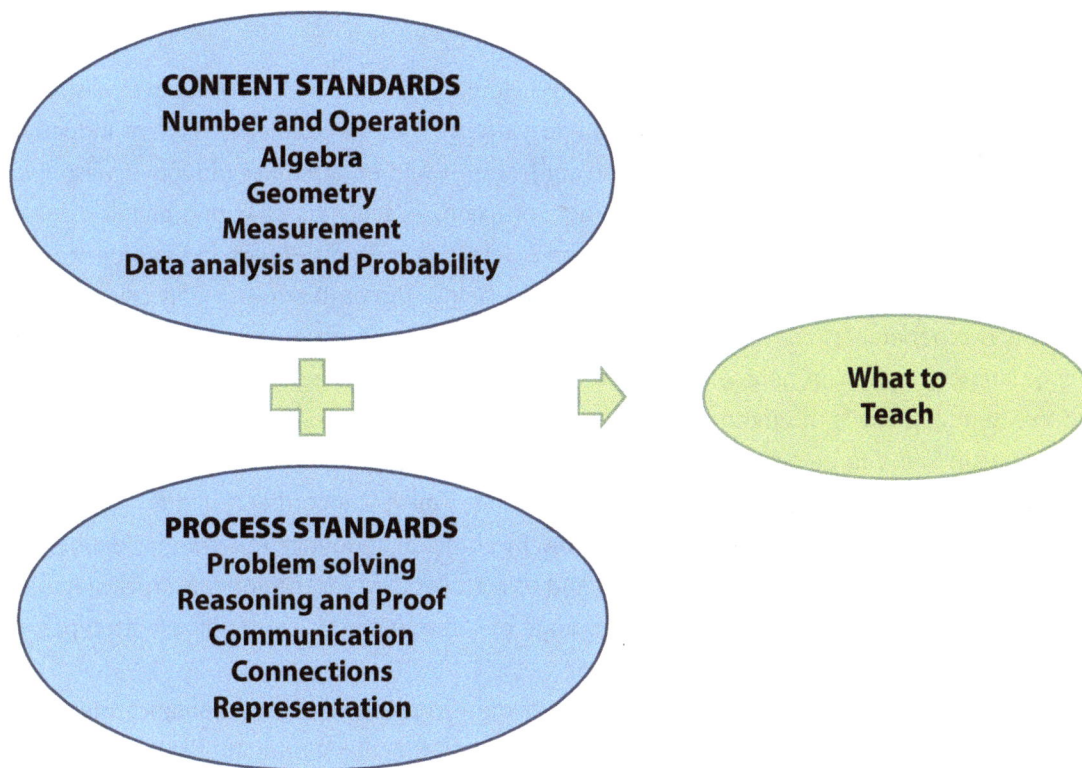

FIGURE 1-1 Basketry-covered Glass Bottle

Source: Executive Summary: Principles and Standards for School Mathematics, National Council of Teachers of Mathematics (2015)

1.3   Process Standards

NCTM (2000) presents five process standards and emphasizes their importance, strongly recommending that children need to practice them in learning mathematics.

Problem Solving

Problem solving is a natural process for young children since the world is new to them. Children solve problems on a daily basis. You need to promote children's ability to pose questions and solve problems by reflecting on their previous experiences and their own ideas. As children grow, problem-solving skills are the cornerstone for formal/school mathematics. According to the NCTM, children in grades pre-K through 12 should be able to do the following:

- Build new mathematical knowledge through problem solving;
- Solve problems that arise in mathematics and in other contexts;
- Apply and adapt a variety of appropriate strategies to solve problems;
- Monitor and reflect on the process of mathematical problem solving.

Reasoning and Proof

There are two types of reasoning: inductive and deductive. Human beings often use inductive reasoning to make the conjecture (or claim), and being able to find patterns is the fundamental skill for inductive reasoning—the process of using patterns to reach the conjecture. Children often use inductive reasoning and become excited by practicing inductive reasoning. For example, teachers of young children use a detective method, such as having children pretend to be a detective to find the patterns in order to get an answer. This allows children to practice their inductive reasoning. The problem with only using inductive reasoning is overgeneralization. Children might identify patterns based on their own experiences and knowledge and come up with generalizations without proof. Children's inductive reasoning is often correct, and they tend to believe this method is flawless. However, this can mislead them: a child sees that numbers plus one are found to be odd numbers (e.g., 2+1, 4+1, 8+1, etc.). The conjectures that the child reaches are that any number plus one is always an odd number, which is incorrect. Children must learn that using only inductive reasoning leads to wrong claims. By contrast, deductive reasoning is defined as a reasoning process starting from a general statement and using logic to reach the conclusion/conjecture. Teachers of young children are recommended to encourage children to practice and utilize both types of reasoning in doing math.

According to NCTM, reasoning and proof skills allow children to further their mathematical thinking. Children with good skills of reasoning and proof can find patterns, functions, or relationships and eventually understand mathematics better. The NCTM presents the following goals to be accomplished by children from pre-K through grade 12:

- Recognize reasoning and proof as fundamental aspects of mathematics;
- Make and investigate mathematical conjectures;
- Develop and evaluate mathematical arguments and proofs;
- Select and use various types of reasoning and methods of proof.

Communication

Communication is an essential part of mathematics education. What do you remember when recalling your mathematics class in your school years? Most likely, your teacher gave you directions and you spent lots of time working on worksheets. It is a frequent observation that mathematics classrooms are very quiet as children work on worksheets without communicating. However, according to empirical studies, math communications improve children's problem-solving skills (Kostos and Shin, 2010) and promote their conceptual understanding of math (Holyes, 1985). This further helps children self-correct misconceptions about math concepts (Kinman, 2010). It has been claimed that communication has been disregarded in early and elementary education math classrooms (Whitin and Whitin, 2002) and should be brought back to them (Kostos and Shin, 2010; Lee, 2014, 2015). Lee strongly recommends strategies such as think-aloud, reasoning and proof questions, and questioning back to children (e.g., use the same wording and ask back to children, "Why do you think?") for use in early and elementary math classrooms to promote math communications.

Communication in mathematics classrooms is critical as children share their mathematical ideas and present what they understand and are able to do. Communication can occur through oral or written communication, symbols, or other representational tools such as drawing, concrete materials, and so on. The NCTM (2000) presents the following goals for children pre-K through grade 12:

- Organize and consolidate mathematical thinking through communication;
- Communicate mathematical thinking coherently and clearly to peers, teachers, and others;
- Analyze and evaluate the mathematical thinking and strategies of others;
- Use the language of mathematics to represent mathematical ideas precisely.

Connection

Connection is the process whereby children are able to connect their mathematical ideas and thinking with other subject areas, their own interests, and their own experiences. Mathematics is not a separate subject, but a pragmatic subject integrated into various areas. This connection process helps children be able to connect their understanding and knowledge of mathematics with new mathematical concepts. The NCTM (2000) requires all children from pre-K through grade 12 to be able to:

- Recognize and use connections among mathematical ideas;
- Understand how mathematical ideas interconnect and build on one another to produce a coherent whole;
- Recognize and apply mathematics in contexts outside of mathematics.

Representation

Children should be able to present and express how they understand mathematical ideas and use these ideas. The term "representation" comprises two aspects: products/outcomes and processes. For younger children, more concrete ways to represent their mathematical thoughts should be accepted and encouraged. As children mature, they can represent their thoughts in a more abstract manner involving oral and written representations using formal terms, symbols, and so forth. The NCTM (2000) standards for children from pre-K through grade 12 include the abilities to:

- Create and use representations to organize, record, and communicate mathematical ideas;
- Select, apply, and translate among mathematical representations to solve problems;
- Use representations to model and interpret physical, social, and mathematical phenomena.

1.4   Common Core Mathematics Standards (Content and Math Practice)

In 2007, the lack of standardization led the Council of Chief State School Officers to discuss the need to develop common standards for the United States and its Annual Policy Forum in Columbus, Ohio. The **Common Core Standards** were presented in 2009. Two major development processes were involved in developing common standards: (1) developing the college- and career-readiness standards (CCRS), which address what students should know and be able to do by the time they graduate from high school, and (2) incorporating CCRS into the K-12 standards in the final version of the Common Core Standards, which address the expectations from elementary school to high school (National Governors Association Center for Best Practices and Council of Chief State School Officers, 2010). Common Core Math Standards (CCMS) include two sets of standards: eight standards for mathematical practice and eleven domains of standards for mathematical contents.

The CCMS's standards for mathematical practice include the following: make sense of problems and persevere in solving them, reason abstractly and quantitatively, construct viable arguments and critique the reasoning of others, model with mathematics, use appropriate tools strategically, attend to precision, and look for and express regularity in repeated reasoning. Standards for mathematical practice address the ways students practice math throughout the school years. CCMS's standards for math contents include eleven domains: counting and cardinality, operations and algebraic thinking (algebra included at high school level), number and operations in base ten, number and operations—fractions (number and quantity included at high school level), measurement and data, geometry, ratios and proportional relationships, the number system, expressions and equations, functions, and statistics and probability (modeling included at high school level). See the table below for standards for contents by each grade level to see the vertical alignment of Common Core Math Content Standards.

TABLE 1-1. *Common Core Math Standards by Content Type*

Content	Grade									
	K	1	2	3	4	5	6	7	8	High School
Counting and Cardinality	X									
Operations and Algebraic Thinking	X	X	X	X	X	X				
Algebra (H)										X
Number and Operations in Base Ten	X	X	X	X	X	X				
Numbers and Operations—Fractions				X	X	X				
Number and Quantity (H)										X
Measurement and Data	X	X	X	X	X	X				

	K	1	2	3	4	5	6	7	8	HS
Geometry	X	X	X	X	X	X	X	X	X	X
Ratios and Proportional Relationship							X	X		
The Number System							X	X	X	
Expressions and Equations							X	X	X	
Functions									X	X
Statistics and Probability							X	X	X	X
Modeling (H)										X

Note. (H) indicates contents apply to high school only.

Data source: http://www.corestandards.org/wp-content/uploads/Math_Standards1.pdf

The Common Core Math Standards also present mathematical practices for children from kindergarten through high school. These practices are closely aligned with NCTM's process standards.

TABLE 1-2. Alignment of NCTM's Process Standards and Common Core Mathematical Practices

NCTM's Process Standards	Common Core Mathematical Practices
• Problem Solving	• Make sense of problems and persevere in solving them
• Reasoning and Proof • Communication	• Reason abstractly and quantitatively • Construct viable arguments and critique the reasoning of others
• Representation	• Model with mathematics
• Representation	• Use appropriate tools strategically
• Problem Solving	• Attend to precision
• Connection	• Look for and make use of structure
• Reasoning and Proof • Communication	• Look for and express regularity in repeated reasoning

Data source: http://www.corestandards.org/wp-content/uploads/Math_Standards1.pdf

1.5 NCTM's Principles for School Mathematics for Young Children

Principles and Standards for School Mathematics also presents six fundamental components to implement high-quality mathematics education: equity, curriculum, teaching, learning, assessment, and technology. These six principles make up the largest part of how to teach young children mathematics and should be reflected in mathematics teaching in early and elementary education (see Figure 1-2).

Equity

The NCTM states that "excellence in mathematics education requires equity—high expectations and strong support for all students" (NCTM, 2000, p. 11). The principle of equity does not imply that there will be the same expectations and the same supports for all children, but that there will be equitable opportunities and high expectations for all children in terms of teaching and learning mathematics.

For example, English Language Learners (ELL) need more support compared to their counterparts who are native English speakers to understand terms and questions in order to successfully solve mathematics problems. Teachers who have ELLs in their classrooms need to consider various supports that will help them solve mathematics problems and learn mathematics with understanding. A gifted child also needs strong support to advance his or her mathematics knowledge and skills. Children with a high level of mathematics skills are sometimes disregarded in a classroom setting. However, close attention to these children is necessary to meet the individual needs of all children. Regardless of whom they are or what their abilities may be, high expectations and strong supports for all children are essential.

Curriculum

The curriculum principle addresses the need for a well-articulated mathematics curriculum. According to the NCTM (2000), "curriculum is more than a collection of activities: it must be coherent, focused on important mathematics, and well-articulated across the grades" (p. 11).

Children from pre-K through grade 12 have the same mathematics content standards but more complex and comprehensive expectations as they mature. For example, children from pre-K through grade 12 learn "Algebra." However, in pre-K through grade 2, children learn about patterns and relationships among physical objects, while children in upper elementary grades learn about more abstract forms involving symbols or formulas.

Teaching

According to the NCTM (2000), "effective mathematics teaching requires understanding what students know and need to learn and then challenging and supporting them to learn it well" (p. 11). The first step in teaching more efficiently is to evaluate what children already know about the topic. To know what children understand about the topic, teachers of mathematics should authentically assess to determine the level of children's knowledge and skills of mathematics (detailed information presented in Chapter 10). The next step is to know what they need to learn. National, state, and local curricula provide resources on what children need to learn. Teachers of mathematics need to be sure that what they teach contains important mathematics. Once teachers evaluate what children know and what to teach, the most important step is how to effectively challenge and support children to learn mathematics with understanding.

Learning

According to the NCTM (2000), "students must learn mathematics with understanding, actively building new knowledge from experience and prior knowledge" (p. 11). In order to do this, children must be able to build new mathematics knowledge from their previous experiences and knowledge. To successfully implement this principle as a teacher, you should be familiar with effective teaching strategies as well as learning theories associated with mathematics learning (see Chapter 3).

Assessment

Assessment should support the learning of important mathematics and furnish useful information to both teachers and students (NCTM, 2000). Assessment is necessary for both teaching and learning to evaluate whether children have mastered target mathematics skills with understanding and whether teachers have efficiently provided mathematics learning opportunities to children.

Assessment is an ongoing process in teaching and learning, and there are a variety of assessments that can be used to authentically assess the quality of both teaching and learning (see Chapter 11 for detailed information about assessments).

Technology

"Technology is essential in teaching and learning mathematics; it influences the mathematics that is taught and enhances students' learning" (NCTM, 2000, p. 11). As technology becomes an integral part of society, children are exposed to it on a daily basis. Technology influences both the ways children learn mathematics and what children need to learn about mathematics.

Utilizing technology facilitates efficient teaching. Sometimes, teachers are concerned that using technology (e.g., calculator) might reduce the opportunity for children to practice mathematics skills. However, using technology tremendously benefits children as they practice mathematics skills by playing with virtual mathematics manipulatives. As you integrate hands-on activities or concrete materials in your teaching, you need to consider technology as a tool.

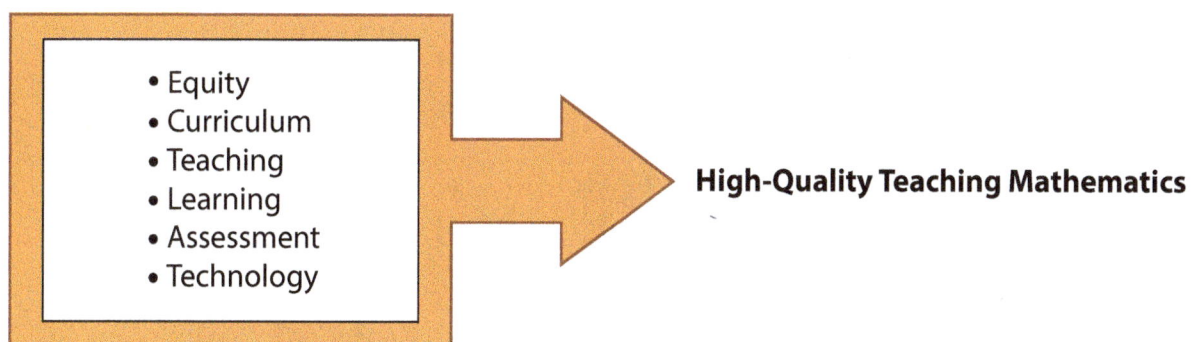

- Equity
- Curriculum
- Teaching
- Learning
- Assessment
- Technology

High-Quality Teaching Mathematics

FIGURE 1-2 NCTM's Six Principles in Teaching Mathematics

Source: Executive Summary: Principles and Standards for School Mathematics, National Council of Teachers of Mathematics (2015)

Reflection Note:

During this week and next week, arrange with an elementary school teacher to observe a math lesson (pre-K through grade 3) and reflect on how the teacher teaches mathematics to young children, focusing on "teaching and learning" principles. In your reflection, include grade level, target mathematics concept(s), and materials the teacher integrated during the lesson.

Levels of Representation in Teaching Children Mathematics

At the end of Chapter 2, you should be able to:

- Describe the levels of representation in teaching children mathematics;
- Explain why you have to use different types of representation.

2.1   What Do Children Bring to the Classroom?

Young children come to school with tremendous informal knowledge of mathematics obtained from their everyday lives, but this knowledge is often underestimated by their teachers, who tend to disregard children's previous experiences and informal knowledge of mathematics by placing more emphasis on teaching formal school mathematics such as terms, symbols, and formulas.

For example, children already have some understanding of the different geometric shapes even though they might not be able to use formal terms to label the shapes. Young children might say, "That is a diamond shape" (indicating rhombus) or "This looks like a sun" (pointing to a circle). Such comments are important indicators of children's knowledge that a teacher should take into consideration when teaching children geometry. Another example of informal mathematics knowledge that is frequently seen in early childhood classrooms is that most children understand how to equally share cookies or snacks with their friends. This is foundational knowledge and skills for understanding division and fractions. You might be surprised to see how precisely young children share snacks or cookies with others by dividing them equally. By connecting what children know with what they need to learn, you as the teacher are able to make mathematics learning more meaningful for them. In teaching mathematics at the early and elementary levels, the most critical task for teachers is to being able to connect children's informal knowledge to formal knowledge, that is, school mathematics (Copley, 2001; Lee, 2014).

How can you, the teacher, connect children's informal knowledge with formal school mathematics? You need to carefully consider the types of representations used in teaching and learning mathematics. Teachers of young children use a variety of materials, including concrete and abstract materials, to teach math. Representation in teaching refers to the ways teachers present math to children. Representation in learning refers to the ways children present their mathematical thoughts and ideas.

2.2 Connecting Children's Informal Knowledge of Mathematics to School Mathematics

The characteristics of cognitive development in early childhood and elementary-aged children allow manipulative materials and models to play a critical role in children's learning mathematics. In order for children to be able to successfully connect their informal knowledge of mathematics to formal or school mathematics, teachers of young children must utilize items that are familiar to them (e.g., concrete materials, pictures, etc.). This principle can be successfully implemented by appropriately providing children concrete, semi-concrete, semi-abstract, and abstract materials (see Figure 2-1). There are three basic principles of using materials that early childhood and elementary teachers of mathematics should consider.

Concrete Representations

Concrete representations can be defined as the ways children present their thoughts using concrete materials (e.g., mathematics manipulatives, physical objects, etc.). When children enter pre-K or kindergarten, they bring tremendous math knowledge to class. In teaching, it is necessary to consider concrete mathematics materials/manipulatives that are familiar to children. For example, providing preschoolers with familiar concrete materials (e.g., toy cars) instead of using commercialized mathematics manipulatives (e.g., color chips, abacus) as counters would help children feel comfortable manipulating objects. Gradually exposing school mathematics to children based on children's previous experiences or familiar items allows them to connect informal and formal knowledge of math. For example, having children play with toy cars for a while and asking them how many cars they have is more comfortable and natural than asking them to count number chips. Encouraging children to count the toy cars by touching or moving them helps them acquire important counting principles such as one-to-one correspondence and cardinality rules (knowing the last number represents the quantity of a set/sets).

Semi-Concrete and Semi-Abstract Representation

Teachers of young children can bridge children's informal knowledge and formal knowledge of mathematics by appropriately utilizing semi-concrete and semi-abstract materials. Sometimes, children become confused when they use toy cars as counters and then are suddenly asked to use written numbers for adding. It is important to naturally connect concrete ways of thinking and abstract ways of thinking by integrating appropriate semi-concrete and semi-abstract materials. Both semi-concrete and semi-abstract representations fall between concrete and abstract representation. Semi-concrete representation is defined as pictorial representation of real objects or materials, while semi-abstract representation uses visual representation that is not associated with real objects. You as a teacher need to assist children to successfully move from semi-concrete to semi-abstract and finally to the abstract level of representation. For example, a child first uses toy cars—a concrete material (representation)—as counters (or as a model). Later, the child may count pictures of toy cars—a semi-concrete representation that represents the real object but is not a concrete material. Next, the child may use tally marks or dots to count; these tally marks function as counters but do not represent real objects. Both concrete and semi-concrete/semi-abstract representations can be used simultaneously to help children bridge concrete and abstract representations.

Using Abstract Representation

Abstract representation can be identified as the terms used in school or formal mathematics such as numbers, symbols, graphs, math terms, and so forth. Abstract representations may make it challenging for children to understand school mathematics. They should be introduced after children have experience with manipulating both concrete and semi-concrete/semi-abstract materials and have a conceptual understanding of particular math concept.

It is important to note that integrating concrete representations is not a must for all children. Some children might be able to understand abstract representation without having experiences with concrete representation. If this is the case, it is not necessary for you to provide those children with concrete materials.

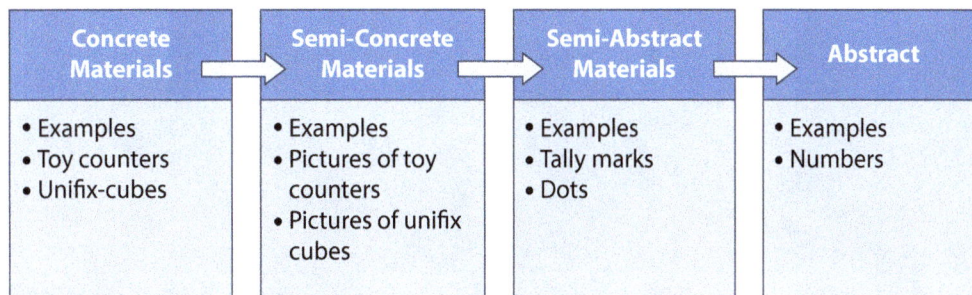

Concrete Materials	Semi-Concrete Materials	Semi-Abstract Materials	Abstract
• Examples • Toy counters • Unifix-cubes	• Examples • Pictures of toy counters • Pictures of unifix cubes	• Examples • Tally marks • Dots	• Examples • Numbers

FIGURE 2-1 Examples of Types/Levels of Representation

Reflection Note:

Observe an early childhood teacher teaching math. Reflect on how the teacher uses different types of representation to teach children mathematics. Also, list the types of representations the teacher utilized. Do you have any suggestions or recommendations for how to improve her/his teaching, focusing on the types of representations? Consider representations from the perspectives of both teaching and learning.

How Children Learn Mathematics

At the end of Chapter 3, you should be able to:

- Differentiate between behaviorist and constructivist approaches in teaching mathematics;
- Describe effective strategies for teaching children mathematics;
- Identify Piaget's types of knowledge and explain why it is important to differentiate types of knowledge in teaching mathematics to young children.

Knowing how children learn mathematics and being able to identify types of mathematics knowledge (physical, logico-mathematical, and social knowledge) facilitates teaching. This chapter addresses theories of learning in early and elementary education to help you understand how children learn and how mathematics should be taught in early childhood education. The most widely used theories on learning in early childhood have two major threads: behaviorist approaches and cognitive, or constructivist, approaches. This chapter presents an overview of how these theories are utilized in teaching and learning mathematics in early childhood and elementary education and Piaget's types of knowledge.

HOW CHILDREN LEARN

3.1 Behaviorist Approaches

In the early twentieth century, education theory was influenced by concepts related to behaviorism. E. L. Thorndike was a major advocate of stimulus and response (S-R) theory. B.F. Skinner furthered this principle and applied it to drill practicing, and it has been widely utilized in the field of education. Teachers or parents with a behaviorist philosophy believe that they can use reinforcement methods to mold children's ideas and learning, essentially making the child become whoever they want him or her to become. If they want the child to become a doctor, by using appropriate forms of reinforcement they can make the child become a doctor. In the behaviorist approach, the child is seen as a passive receptacle for knowledge.

Some examples of positive reinforcement that early and elementary teachers use include giving out stickers, candies, or good scores for those children who earn expected scores or high scores. Negative reinforcements include such techniques as time-out, punishments, any types of negative remarks on children's report cards, and so on. Negative reinforcement can have immediate impacts on children's behaviors by reducing undesirable behaviors or increasing desired behaviors, but it should not be frequently applied. Negative reinforcement often causes negative feelings in children.

3.2 Cognitive or Constructivist Approaches

A more recent theory about learning is the cognitive/constructivist approach, which is based on the principle that children are able to actively construct their knowledge. Instead of directly teaching mathematics concepts to children, you as a teacher facilitate children's learning by providing appropriate materials and environment. In these types of approaches, children should play an active role in acquiring knowledge rather than receiving it in a passive manner.

Piaget's Cognitive Theory

The major advocate of cognitive/constructivist approaches was Jean Piaget, a Swiss philosopher and epistemologist. The central concept of Piaget's theory is that children's mental structures result from operations developing through various stages. Piaget presents four stages of cognitive development: sensorimotor, preoperational, concrete operational, and formal operational. Most children in early and elementary levels fall into the first three categories. Major characteristics of children in the preoperational stage are egocentric thinking and an inability to reverse events (irreversibility), which is associated with the concept of conservation. Children in both preoperational and concrete operational stages have not yet developed abstract thinking ability and need to be provided with concrete materials to manipulate in order to learn mathematics.

Child-Centered Mathematics Learning

Child-centered mathematics learning is a representative example of the constructivist approach. It emphasizes children's learning more than the teacher's teaching. The roles of teachers using this approach are to provide a rich mathematics environment and materials, offer important mathematics tasks, ask about the rationales of children's responses or answers, listen actively, and so on.

A very popular type of child-centered math learning in early childhood classrooms is "center time," which is often included in everyday routines. Math centers are set up to present various materials based on math curriculum or standards. Children freely manipulate mathematics materials at the center during center time or free-play time. Providing children with center-based time to learn and manipulate mathematics materials helps them learn mathematics at their own pace. A teacher needs to make sure that all children are learning mathematics with understanding at their ability.

Cooperative Learning

Group work, sometimes called "collaborative learning," has been emphasized by mathematics and early childhood educators and researchers. Placing children into groups of three or four to work on mathematics problems is an important teaching strategy that enables children to communicate and interact with their peers mathematically. By sharing and explaining mathematics ideas with their peers, children are able to clarify their thinking as well as listen to their peers' mathematical ideas. You as the teacher need to consider how to group children, such as by heterogeneity of skills.

Table 3-1 presents the characteristics of the behaviorist approach versus the constructivist approach in education.

TABLE 3-1 Behaviorist vs. Constructivist Approach in Mathematics Learning and Teaching

	Behaviorist Approach	Constructivist Approach
Advocate	B. F. Skinner	J. Piaget
Principles	Reinforcement/Stimulations	Internal motivation, children constructing their own knowledge
Teacher's Roles	Providing children appropriate reinforcement, either positive or negative, to maximize their mathematics learning	Providing children appropriate materials and environment to facilitate their mathematics learning
Child	Passive existence	Active existence
Examples	Awards, scores or grades, stickers, compliments, time-out, punishment, removing recess, etc.	Using daily mathematics journal to reflect on what they learn and know, mathematics center, mathematics games or play, etc.

3.3 Recommended Strategies in Teaching Children Mathematics

Some children learn better when they actually manipulate materials, while others learn more efficiently as they work with others. It is not easy to define the most effective strategies for all children in teaching mathematics, but some strategies have been shown to be more effective than others. Some you might want to consider include:

- Actively engaging children in learning mathematics
- Starting from what children know
- Providing concrete materials to promote children's understanding of mathematics

- Utilizing questioning strategies
- Applying grouping strategies
- Connecting mathematics with real-life experiences
- Promoting mathematics communications

PIAGET'S TYPES OF KNOWLEDGE

Jean Piaget differentiated knowledge into three types of knowledge, which are physical knowledge, logico (logical)-mathematical knowledge, and social knowledge. The following presents how Piaget's types of knowledge are used in mathematics.

3.4   Physical Knowledge

Physical knowledge is the very first knowledge children acquire by observing their surroundings and playing with toys or other physical materials. This includes knowledge involving how physical materials or objects appear as a result of their attributes or characteristics. Physical knowledge doesn't require any social agents (parents, guardians, teachers, or peers) to acquire and doesn't require knowledge of relationships between physical objects. This is considered as knowledge "out there" in the physical world without manipulating. An example of physical knowledge in mathematics is being able to collect objects by characteristics or attributes without involving names or terms of attributes of shapes or colors.

3.5   Logico (Logical)-Mathematical Knowledge

Logico-mathematical knowledge is based on patterns and relationships between objects. Unlike physical knowledge, children should process information by manipulating or observing manipulated objects to acquire logico-mathematical knowledge. Based on observations, children tend to relate or to pattern their observations of physical objects based on specific attributes, which is referred to logico-mathematical knowledge. This can't be told. If someone explains about relationships between two objects, this is considered as social knowledge. For example, children should learn what numbers are odd and what numbers are even, which is referred to as social knowledge. A teacher will tell kindergarteners 1, 3, 5, 7, and 9 for odd numbers and 2, 4, 6, 8, and 10 for even numbers. This is considered as social knowledge. However, once children find out number patterns by themselves and are able to identify even or odd numbers out of any numbers based on number relationship and patterns without being told, it is called as "logico-mathematical knowledge based on deductive thinking process." Logico-mathematical knowledge can be built in either the deductive thinking process or inductive thinking process. When children find out relationships and patterns between and among physical objects and generalize the relationships/patterns, this is considered as logico-mathematical knowledge based on the inductive thinking process. Children should be the only entity to build and construct logico-mathematical knowledge.

3.6   Social Knowledge

To obtain social knowledge, it is essential to have social agents or social media (TV, Internet, books, etc.) to deliver information to children. Social knowledge should be given by social agents or social media. Children learn social knowledge in social and cultural contexts through social agents. For example, being able to count from one to ten is considered social knowledge since children can't learn to say counting numbers by themselves without social agents or media. Knowledge associated with language and culture is also considered as social knowledge (e.g., being able to communicate in a certain language, greeting manners, social manners, etc.). In mathematics, examples of social knowledge are "converting measuring units," "knowing terms along with definition" (e.g., triangle, rectangle, polygon, etc.), "knowing formula," etc. Things that should be learned from others are considered as social knowledge.

Reflection Note:

Observe an early childhood teacher teach mathematics and reflect on whether the teacher applies behaviorist or constructivist (cognitive) approaches in his or her instructional methods. In addition, observe how the teacher differentiates teaching methods when teaching children math based on the types of Piaget's knowledge.

Number and Operations I:

Developing Early Number Sense in Early Childhood

At the end of Chapter 4, you should be able to:

- Describe pre-number concepts;
- Describe how "subitizing" promotes children's number sense.

Table 4-1 presents NCTM's standards on Number and Operations and behavioral expectations for children from pre-K through grade 2. In addition, a Common Core Math Standards overview is presented: "counting and cardinality" for kindergarten and "number and operations in base ten" for kindergarten through grade 2.

TABLE 4-1 NCTM's Number and Operations Standards for Pre-K through Grade 2 & Common Core Math Standards for K through Grade 2

NCTM's Number and Operations (NCTM, 2000, p. 78)		Common Core Math Standards Overview on Number and Operations (National Governors Association Center for Best Practice & Council of Chief State School Officers, 2010)
Number and Operations Standards for pre-K through Grade 2	Behavioral Expectations	**Grade K (10)** **Counting and Cardinality**
• Understand numbers, ways of representing numbers, relationships among numbers, and number systems	• Count with understanding and recognize "how many" in sets of objects • Use multiple models to develop initial understandings of place value and the base-ten number system	• Know number names and the count sequence. • Count to tell the number of objects. • Compare numbers

(Continued)

NCTM's Number and Operations (NCTM, 2000, p. 78)		Common Core Math Standards Overview on Number and Operations (National Governors Association Center for Best Practice & Council of Chief State School Officers, 2010)
	• Develop understanding of the relative position and magnitude of whole numbers and of ordinal and cardinal numbers and their connections • Develop a sense of whole numbers and represent and use them in flexible ways, including relating, composing, and decomposing numbers • Connect number words and numbers to the quantities they represent using various physical models and representations • Understand and represent commonly used fractions, such as ¼, ⅓, and ½	**Number and Operations in Base Ten** • Work with numbers 11-19 to gain foundations for place value **Grade 1 (p.14)** **Number and Operations in Base Ten** • Extend the counting sequence • Understand place value • Use place value understanding and properties of operations to add and subtract. Grade 2 (p.18) **Number and Operations in Base Ten** • Understand place value. • Use place value understanding and properties of operations to add and subtract.
• Understand meanings of operations and how they relate to one another	• Understand various meanings of addition and subtraction of whole numbers and the relationship between the two operations • Understand the effects of adding and subtracting whole numbers • Understand situations that entail multiplication and division, such as equal groupings of objects and sharing equally	
• Compute fluently and make reasonable estimates	• Develop and use strategies for whole-number computations, with a focus on addition and subtraction • Develop fluency with basic number combinations for addition and subtraction • Use variety of methods and tools to compute, including objects, mental computations, estimation, paper and pencil, and calculators	

The National Association for the Education of Young Children (NAEYC) and National Council of Teachers of Mathematics (NCTM) have emphasized the critical importance of early childhood mathematics education because it has a great impact on children's later mathematics achievement (NAEYC and NCTM, 2002). Young children should have a strong foundation of number sense on which to build more advanced mathematics concepts and skills.

4.1 Pre-Number Concepts and Numbers in Early Childhood

Pre-number concepts are the early skills that are not directly associated with numbers but that help children develop their later number sense. These skills include classification and recognizing patterns.

Classification

Children classify physical objects in their everyday real life. Even infants or newborn babies are able to classify by showing different reactions to unfamiliar faces from familiar faces. For example, an infant might cry when she sees a stranger and recognizes her mom by smiling at her mom. Classification skills are fundamental to children's number sense. The ability to classify objects allows children to differentiate geometric shapes, types of attributes, various sizes, various amounts, and so forth. Being able to classify objects allows children to know what to count. If children are asked to count the blue marbles in a group of different colored marbles, they should first be able to classify the blue marbles in the set. If a teacher asks children how many boys are in the class, children first should be able to classify boys and girls to successfully perform the task. Classification skills are also essential for all types of sorting tasks that frequently appear in mathematics problems and are fundamental for algebraic thinking and algebra, which require the student to sort objects by patterns and to find patterns to solve problems. Therefore, teachers of young children should be alert and ready to identify and offer activities to help children develop their classification skills.

Patterns

Mathematics can be defined as the study of patterns. Identifying, describing, and creating patterns are essential activities that help young children develop their number sense, especially for algebraic thinking. Without patterns, math wouldn't exist. Providing various opportunities for children to manipulate materials and do pattern-associated activities is necessary in early childhood mathematics. For example, having children copy given patterns or shapes and extend patterns helps them to develop their algebraic thinking. Encouraging children to copy shapes (pattern block shapes, geo-board shapes, tangram shapes, etc.) helps them concretely grasp the concept of patterns.

4.2 Early Number Concepts in Early Childhood

Some early number concepts are necessary to advance children's number sense to more complex forms. For example, children will not be able to rationally count objects without one-to-one correspondence skills. They will also have limited skills of mental computation and estimation if they cannot recognize groups.

One-to-One Correspondence

One-to-one correspondence is a useful and important strategy in order for children to obtain rational counting skills. In counting, one-to-one correspondence refers to one number representing one object.

FIGURE 4-1 One-to-One Correspondence

When children count, they need to know that one item matches with one number, representing one-to-one correspondence. Without this understanding, they will not be able to count rationally as is required in computation. Children sometimes make a mistake by skip-counting object(s) or counting the same object twice. To help children practice one-to-one correspondence, providing them a small container to use as they count objects would eliminate unnecessary mistakes. When children count pictures of objects, they can cross off each picture as they count. This would reduce any potential mistakes in counting by ensuring children's one-to-one correspondence. It is also recommended to encourage children to use one-to-one correspondence in everyday life by asking a child to distribute a piece of paper to each child.

Number Conservation

Conservation concepts originate from Jean Piaget's theoretical framework. According to Piaget, children in the preoperational stage lack understanding of the concept of conservation (e.g., conserving various types of tasks such as number, volume, area, length, etc.) because they do not understand reversibility. For example, in Figure 4-2, there are two rows of chips arranged side-by-side. A teacher asks a five- or six-year-old child to compare the quantities of these two groups (i.e., which row has more chips?). The child may say that each row has the same number of chips. Now, the teacher increases the distance between the chips in the second row in front of the child (child's sight) and asks the child to compare quantities of the two groups. The child may say that the second row has more chips without counting the chips. This implies that the child has not yet developed an understanding of number conservation.

Task 1

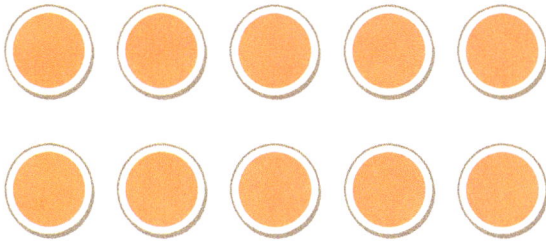

Task 2

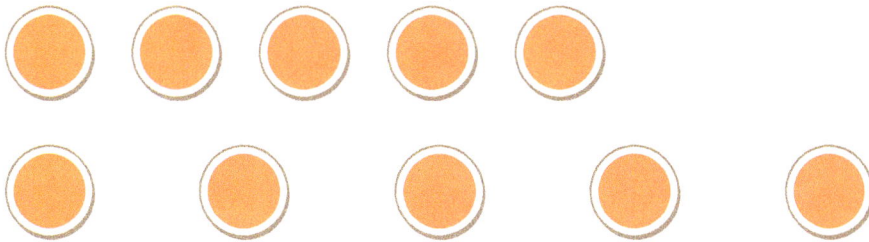

FIGURE 4-2 Conservation Tasks

Group Recognition

Group recognition is another early number concept that promotes children's number sense. By looking at the patterns of objects or things, children can easily compare the quantities of groups. According to Clements (1999), children can easily identify the numbers from one to three without counting before they enter school (e.g., one nose, two eyes, three wheels on a tricycle, etc.). This is a type of group recognition that is a frequent observation in early childhood.

FIGURE 4-3 Group Recognition

One important skill of group recognition is "subitizing." Subitizing is from a Latin term meaning "suddenly"; it is defined in mathematics education as knowing instantly how many are in a set. When children see a familiar pattern like dominoes or dice, they will be able to say how many dots are in a set without counting. A child who is familiar with dominoes will look at Figure 4-3 and be able to instantly say that is "five" without counting the dots.

Strategies of subitizing or sight recognition of quantities help children in many ways. First, they save time. By seeing the group of objects, children will be able to know the quantities of numbers without counting. In more advanced tasks, children will be able to count the numbers from the point where they can count with sight recognition (Figure 4-4). For example, a child might know the quantity ("six") of the first domino card without counting (Figure 4-4) and start counting from "seven" when adding these two domino cards. More importantly, subitizing practice promotes children's number sense by providing children perceptual anchors. For example, when children see Figure 4-4, they will automatically know the numbers of dots in both cards since those are familiar patterns to them. They will immediately count 5 + 4 without counting dots from each card. Sometimes, subitizing functions as mental manipulatives. When children are shown dots on a card, they don't have time to actually count the number of dots. Sometimes, children would say the number of dots immediately. Other times, they often mentally visualize the number of dots and patterns on a card and count them to say the number.

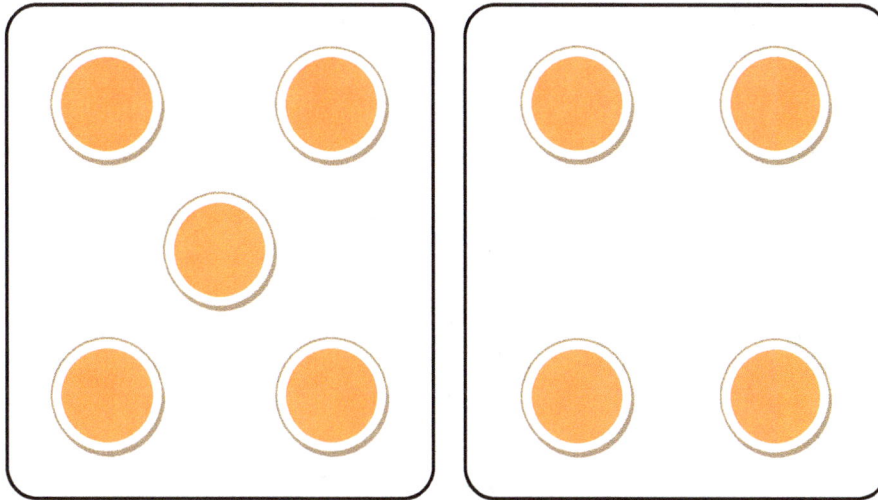

FIGURE 4-4 Subtizing Example 1

Subitizing also helps children compare quantities of objects without counting them. When a child sees Figure 4-5, the child will be able to compare the quantities of two groups. This skill leads children to develop number sense involving the skills of addition and subtraction.

Comparisons

Comparisons are frequently used in all areas of mathematics. Most importantly, the ability to compare quantities is an important early number concept for children to acquire in early mathematics. The ability to make comparisons is an essential skill in early mathematics and becomes the foundation of later number and operations such as addition, subtraction, multiplication, and division.

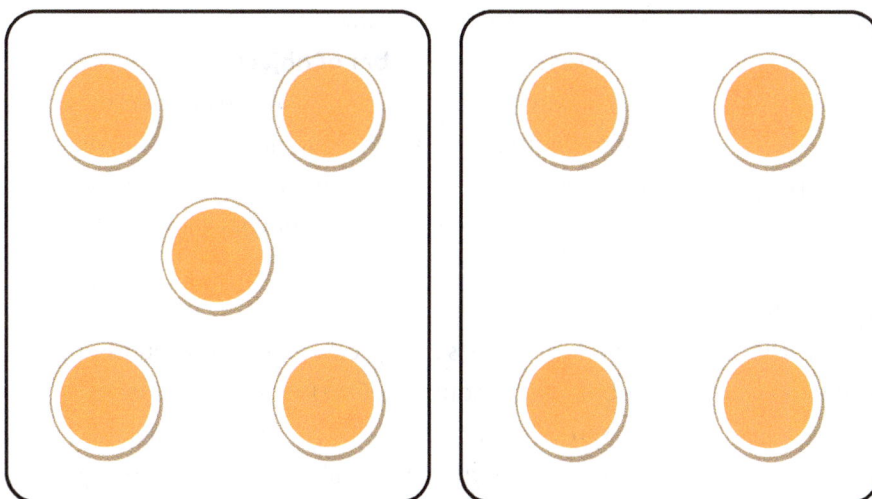

FIGURE 4-5 Subtizing Example 2

4.3   Number Concepts in Early Childhood

In order for children to build number concepts with counting ability, children should obtain the following four counting principles.

Counting Principles

Common Core Math Standards present "counting and cardinality" as the fundamental skills to be obtained during the kindergarten year. Children must understand the following four principles in order to perform rational counting.

A. One-to-one correspondence (see more explanations above)

Children should be able to assign one number name to one object without skipping any objects or without counting any objects more than one time in order to be able to rationally count. Children often skip object(s) when counting. For example, a child may count four chips by saying one, two, four, skipping one chip, or a child may count five chips by saying one, two, three, four, five, six, counting one chip twice. The child may say he has six chips. Both are examples of a child lacking the skills of one-to-one correspondence. In order to help a child enhance this skill, it is necessary to provide a child counters with a number of containers. When children count chips/counters, it is recommended they put one chip/counter into one container in order to practice one-to-one correspondence. Children may move counters from one side to other side when counting.

B. Cardinality rule

The cardinality rule is one of the most important skills in counting. It has been specifically emphasized in Common Core Math Standards (CCMS) as a required skill for kindergarteners (i.e., connecting counting to cardinality). According to CCMS, kindergarteners should be able to answer to "how many?" questions about as many as twenty things arranged in various shapes (Common Core Standards Initiative, n.d.). **Children should know that the last number represents the number of objects**. This rule is associated with "how many" are in a set. Regardless of order, children should be able to say the last number counted when they are asked how many are in a set. To help children to obtain the cardinality rule, providing children concrete materials (counters) is essential. Some examples are any counting activities/games with counters, placing stickers on written numbers, counting objects together, or subitizing practice activities.

C. Order irrelevance rule

When children count, they should know that there is no correct order when counting objects. They might start from the first object or the last object when counting. Children should understand that this does not impact the quantity of objects. To help children obtain the order irrelevance rule, a teacher would provide children a container and counters dispersed over the table. Children may pick up any counter one by one without any rules and count the number of counters when placing into the container.

D. Stable order rule in saying/counting numbers

To be able to successfully perform rational counting, children should first be able to memorize the numbers in a correct sequence without skipping or saying the same number twice (e.g., one, two, three, four…), which is called "rote counting." Rote counting is essential for rational counting. Children MUST say the numbers in the correct order. To do so, teachers of young children should provide various activities utilizing counting songs or counting games on a daily basis to practice saying numbers in a stable sequence. Again, counting numbers from one to ten is a type of social knowledge, which means that teachers of young children should deliver the information in the way children enjoy to learn and to memorize to ultimately help children become competent in counting.

Counting Types

There are two types of counting teachers of young children should consider: rote and rational counting.

A. Rote Counting

Children in the early age range enjoy counting numbers informally. They often say numbers in incorrect order or say the same numbers more than one time. The following examples are common mistakes associated with counting numbers and one-to-one correspondence children make frequently in the stage of rote counting.

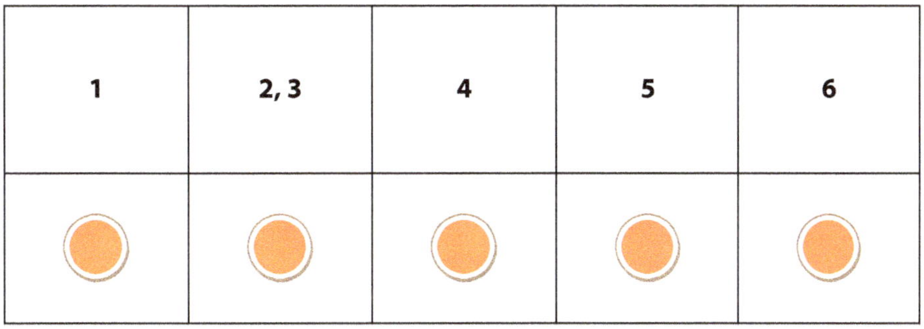

1	2, 3	4	5	6
●	●	●	●	●

FIGURE 4-6 Counting Too Fast

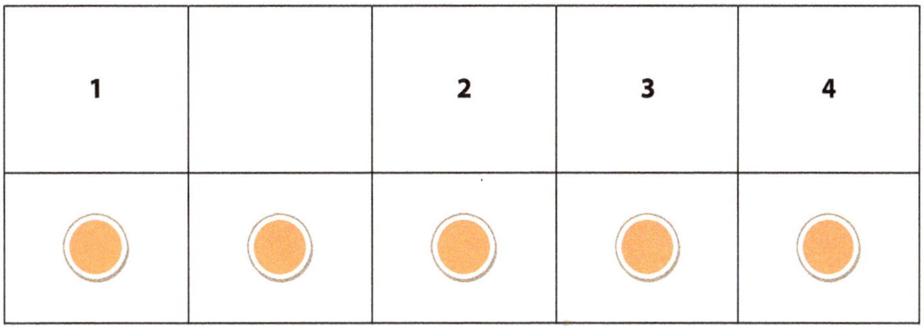

1		2	3	4
●	●	●	●	●

FIGURE 4-7 Counting by Skippling Some of the Objects

Incorrect sequence of counting and correct one-to-one correspondence

When children count objects, they may use a one-to-one correspondence strategy, pointing to each object one at a time. However, they might say numbers in an incorrect manner. For example, a child may say, "One, two, three, five, six, eight, ten." In this example, a child may not be able to say the correct number of objects.

Correct sequence of counting and incorrect one-to-one correspondence

This is the opposite case. Children say numbers in the correct sequence, but they might use one-to-one correspondence in an incorrect way. For example, a child may count objects too fast, saying, "One, two, three, four…" correctly but without matching the number to an object, or they might say, "One, two," pointing to just one object. Or a child may say numbers in a correct sequence but point to some objects and skip others. Any of these mistakes will lead to errors in counting.

B. Rational Counting

Children who are able to count rationally can say numbers in a correct sequence using correct one-to-one correspondence strategy and understand cardinality. Rational counting ability is the foundation for the advanced number sense needed for later school mathematics.

According to the NCTM (2000), children from pre-K through grade 2 should be able to rationally count and should be able to identify "how many" in sets of objects. (See the NCTM's expectations for early number sense.)

Reflection Note:

Observe a pre-K or kindergarten classroom and reflect how the teacher enhances children's pre-number concepts and early number concepts.

Number and Operations II:

Developing Children's Number Sense

At the end of Chapter 5, you should be able to:

- Describe four common rules of the Hindu-Arabic numeration system;
- Differentiate between proportional and non-proportional models;
- Describe three types of fraction models.

5.1 Hindu-Arabic Numeration System

In North America, we follow the common rules of the Hindu-Arabic numeration system, which has four major characteristics: place value, base of ten, use of zero, and additive property. Helping children understand these basic principles of the number system eventually helps them to practice mathematics more efficiently. This chapter presents how teachers of young children teach mathematics based on these four principles.

Place Value

Place value means that each place has a different value. This is a challenging concept for young children because quantitative value is determined based on the place value instead of the numeral quantity. For example, let's look at the number 123. Considering the value of each number, 1 has the least quantitative value and 3 has the most quantitative value. However, based on the Hindu-Arabic numeration system, place value is more important to consider than quantitative value. In this case, the 1 is located in the hundreds place and so represents 100. However, 3 is located in the ones place, indicating 3. Thus, 1 representing one hundred has more quantitative value than 3 representing 3 ones.

According to Common Core State Standards for Mathematics, children from kindergarten through grade 3 should work with numbers to understand place value and operations in base ten. Especially, kindergarteners should practice to gain foundations for place value by working with 11 to 19. In order to help children have concrete understanding of place value, it is important to expose children to various proportional materials in which children can visually see 10 ones from 1 ten (e.g., a bundle of ten Popsicle sticks, base-ten blocks, a bag of ten counters, etc.) As children grow, it is necessary to integrate the abstract form of math materials.

Base of Ten

The number system we use is base ten. This means that there cannot be a value higher than nine in each place. If there is more than the quantitative value nine, the number will be traded to the next digit. Furthermore, a base of ten system has ten digits (zero to nine). "Base of ten" is closely related with written numbers. Children need to practice counting concrete materials and connecting to the corresponding written numbers. In addition, children learn to say written numbers. A good activity example to promote children's understanding of "base of ten" is the use of a number frame. When children use number frames, it is recommended to place a counter to the corresponding number. For example, provide children ten-frames and let them place 12 counters on the ten-frames. Encourage them to write the number of counters on the blank paper (12).

Use of Zero

Zero has an important meaning in the system because it represents the "absence of something," which is closely associated with place value. Sometimes, young children are confused about the meaning of zero. They often think that zero means "nothing." This leads children to a problem when they encounter a number containing zero. For example, a child who sees the numerals 101 and 99 might be confused when comparing these two numbers' quantitative values. Children often look at which numeral is made up of numbers with higher quantitative values (e.g., seven to be higher than six, six to be higher than five, etc.) and tend to believe that the number 99 has higher quantitative value than 101. It is important to help children understand the true meaning of zero by referring to "absence of something." For example, let's consider the number 305. In the system, 0 means absence of tens. Another example is the number 1020. The first 0 means absence of hundreds, and the second 0 means absence of ones. Children need practices to match counters to the corresponding written numbers. Either children should be able to write the corresponding numbers by looking at counters/objects (e.g., base-ten blocks, a bundle of Popsicle sticks, etc.) or children should be able to show the materials corresponding to the written numbers. This practice helps children concretely understand the concept of zero in various written numbers.

Additive Property

The Hindu-Arabic numeration system uses the additive property of numbers. This is also observed in an extended form of writing. For example, let's look at the number 342. Based on the additive property rule, this number is composed of 300 + 40 + 2. Every number in the Hindu-Arabic system is affected by the additive property rule. Having children practice breaking numbers by place value (e.g., hundreds, tens, and ones) helps them make sense of the additive property. Knowing principles of the additive property promotes children's number sense.

5.2 Models/Materials to Help Children Enhance Place Value and Base-10 Principle

Based on the place value, every number can have different quantitative value. It sometimes is difficult for young children to understand the concept of place value since they tend to see only one dimension (seeing the quantitative number value instead of seeing place value). For example, let's take the number 25. Young children might see the 2 as 2 ones rather than 2 tens. This leads them to conclude that 5 is a bigger number than 2. So, how does the teacher help children to appropriately understand place value? Modeling (ungrouped vs. pre-grouped and proportional vs. non-proportional) helps children better understand place value along with the concept of "base of ten."

Ungrouped vs. Pre-Grouped

Teachers of young children should provide plenty of opportunities for children to practice counting using ungrouped materials such as beads, toy cars, Legos, and so on. Once children play with ungrouped materials, they can also group the materials following the teacher's directions or on their own. Manipulating the objects promotes quantitative reasoning as well as an understanding of place value. Pre-grouped materials are grouped by certain numbers (e.g., fives or tens). Base-ten blocks are the representative example of pre-grouped materials (e.g., pre-grouped 10 represents ten ones, pre-grouped 100 represents 10 tens). Appropriately using both types of materials benefits children as they develop their number sense and place value.

Proportional vs. Non-Proportional

Proportional models are easier for young children to use in practicing number sense than non-proportional models. When children manipulate proportional models, they can actually see quantities. A base-ten block is a perfect example of a proportional model. One 10 consists of 10 ones, and one 100 consists of 10 tens or 100 ones. One 1000 (a cube) consists of 10 one-hundreds, 100 tens, or 1000 ones. Children can actually see these proportions on the base-ten blocks; one 10 is actually ten times bigger than one 1, and one 100 is a hundred times bigger than one 1 and ten times bigger than ten 10s. A meter stick is another example of a proportional model. In fact, most measuring tools can be categorized as proportional models.

However, non-proportional models are not related to size. Though non-proportional models are not easy for young children, they must learn about them. Non-proportional models are prevalent in upper grade levels in mathematics curriculum and include money units, abacus, binary scales, and so on. Money is probably the most popular example of a non-proportional model. Children cannot see five pennies in one nickel or a hundred pennies in one dollar. Money (non-proportional model) is not a good model for introducing place value since children tend to see size first. A common mistake children make is thinking that a nickel has more value than a dime. Children often see one dimension, either size or quantity, without considering the value of the object. As a result, they are often willing to exchange one dime for five pennies or one quarter for five or ten pennies.

An abacus is another example of a non-proportional model. Each color of beads represents place value. For example, one red bead represents 1 and one green bead represents 10. In this case, children cannot see the comparable quantitative value between one red and one green bead.

Using proportional models is the more appropriate method for promoting young children's number sense or teaching place value. Once children become comfortable with manipulating proportional models, the teacher can naturally integrate non-proportional models.

5.3 Zero as Absence of Something

Children often have a difficult time understanding the meaning of 0. When children see 0, they frequently consider it as a representation of the least quantity; they always list it as the number representing less than one or any other numbers. However, when children are exposed to 0 used in other than the ones place, they become easily confused. To help children understand the meaning of 0, it is necessary to help them understand "0" as the absence of something instead of as the number with the least value. To do so, children must use concrete materials along with written numbers when they begin to learn numbers. As Figure 5-1 presents, it's challenging for young children to see "20" as "2" tens and "0" ones. They might see "20" as "2" and zero. This violates place value as well as additive property. To help children visually see zero as absence of something, it is necessary to By expose them more concrete experiences to practice matching number of objects corresponding to the written numbers and vice versa.

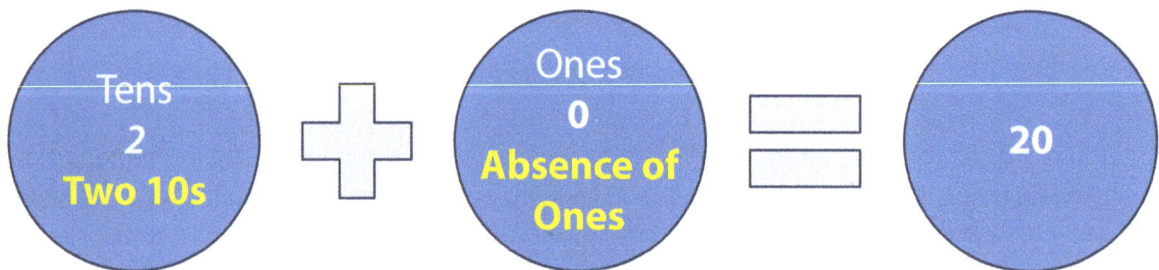

FIGURE 5-1 The Use of Zero

5.4 Additive Property

The additive property rule applies in all of our numeral systems. For example, children need to learn that the number 353 consists of 300 + 50 + 3. Base-ten blocks and number mats are good ways to help children bridge concrete and abstract forms of mathematics when teaching about the additive property.

5.5 Helping Children Develop Number Sense

Develop Number Benchmarks

According to the NCTM (2000), children should be able to show fluency with accuracy when they do mathematics. Helping children use number benchmarks can promote their intuitive number sense. Number benchmarks are defined as "perceptual anchors" that result from children's concrete experiences of doing mathematics. Number benchmarks are often used when children count or add numbers, especially the numbers 5 and 10. For example, when children add 6 + 7, they might start from 6 and take 4 out of 7 to make 10. They look at the remainder 3 out of 7 and come up with 13 as an answer. In this case, children used 10 as their number benchmark.

Provide Concrete to Abstract Materials

As described in Chapter 2, children need appropriate types of representations (from concrete to abstract) to develop their number sense. For example, you need to provide all children accessible counters when they add. Once they feel comfortable with using counters, they can begin to use tally marks. Later, they will do mathematics in an abstract form using symbols, numbers, equations, and so on.

Make Connections

To make mathematics learning more meaningful for children, it is necessary to connect mathematics content with children's real lives. Instead of asking children to solve story problems from a math book, you can actually bring real-life problems to class. For example, tell children that they need a new carpet for the classroom and ask them to find the area of the classroom so that you can order the new carpet.

Regrouping and Trading

Regrouping is a type of trading skill. If you get a sum that is greater than nine, you must regroup/trade because of the base-ten principle. Ten ones should be traded for one ten. However, the term "regroup" is too complex and abstract for children to understand, and misunderstanding of this term leads them to make errors in solving arithmetic problems. It is easier to do regrouping/trading in addition than in subtraction. For this reason, you as the teacher need to provide children concrete experiences with subtraction using regrouping. Figure 5-2 shows how you can teach the regrouping strategy to young children using a concrete model (number mat or number column).

Understanding Properties

The commutative property of numbers means that when two numbers are added, the sum is the same regardless of the order of the addends. For example, 4 + 2 equals 2 + 4. The associative property means that when three or more numbers are added, the sum is the same regardless of the order of addition.

For example, (2 + 3) + 4 equals 2 + (3 + 4). The distributive property means that the number that is outside the parentheses must be distributed to the numbers inside the parentheses. You can multiply the number outside the parentheses by the other numbers on the inside of the parentheses and then add or subtract accordingly. For example, 4 × (3 + 2) equals (4 × 3) + (4 × 2) and 4 × (3 − 2) equals (4 × 3) − (4 × 2).

Problem: 35 − 27 = ?

• The first step is to insert the numbers into columns

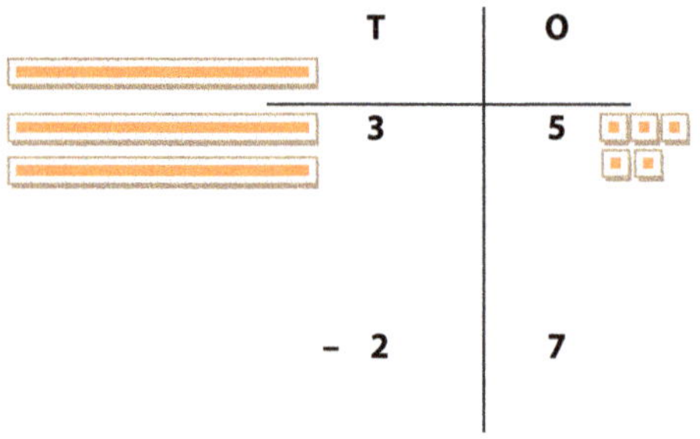

** Note: T = Ten's Place, O = One's Place

• The second step is to regroup in the ten's place.

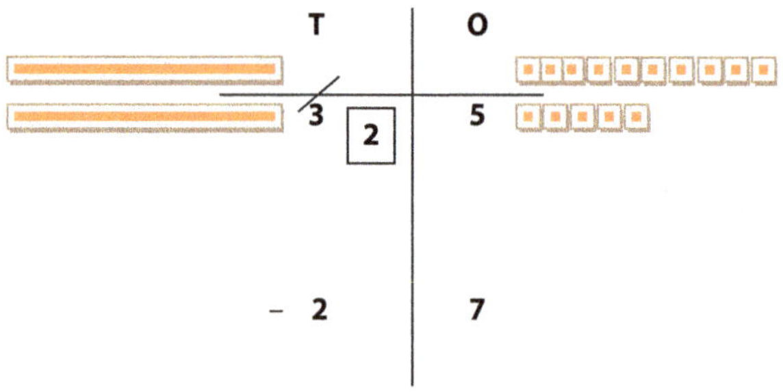

• The third step is to subtract.

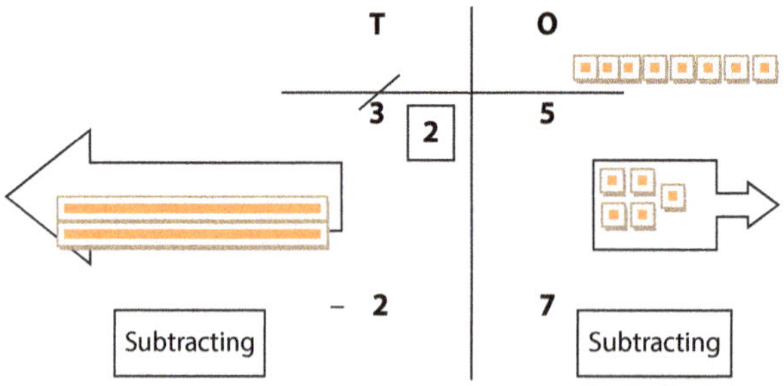

• The final step is to count how many blocks are left. The answer is 8.

FIGURE 5-2 Concrete Examples of Regrouping

5.6   Fractions and Decimals

In grades 3 through 5, children learn about fractions and decimals in a more abstract form. Children in pre-K through grade 2 should have an understanding of basic forms of fractions such as ½, ⅓, ¼, etc. As children move to upper grade levels, they should be able to compare quantitative values of fractions. Fraction towers or fraction strips are useful tools for children to manipulate fractions. Using these materials, children can visually see the size of fractions and physically compare and contrast sizes based on the relationships between fractions. To teach children fractions more efficiently, you need to integrate appropriate models of fractions.

A fraction has three meanings: part and whole, quotient, and ratio. The meaning of part and whole in a fraction is the most often-used concept in early childhood, and the part and whole model refers to a whole that is partitioned into equal parts. The fraction is also regarded as a quotient. For example, ¼ can be considered as a quotient 1 divided by 4. The fraction notation ¼ can also represent a ratio (one to four). For example, a child can spend one hour for outdoor play time for every four hours of inside classroom activities.

Region Model

A region model is the simplest and easiest form of fraction models. The region model is characterized by being made up of shapes that are all the same size. As shown in the example, a chocolate bar can be used as a region model, with the yellow-shaded area representing the fraction 3/15.

Figures of "pizza" or "cookie" are also popular region models. The fractions in these figures must have the same shapes with the same size to represent the fractions.

FIGURE 5-3 Region Model-Example 1 Representing 3/15

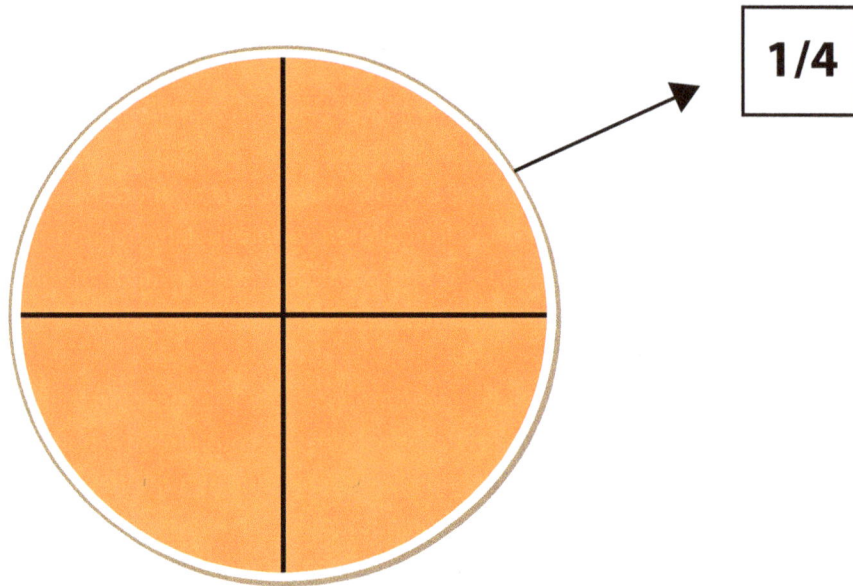

FIGURE 5-4 Region Model-Example 2 Representing ¼

Length Model

A length model is a little more advanced for children to understand. You can use a line or piece of yarn to model fractions. An often-used length model is a number line (see the example below).

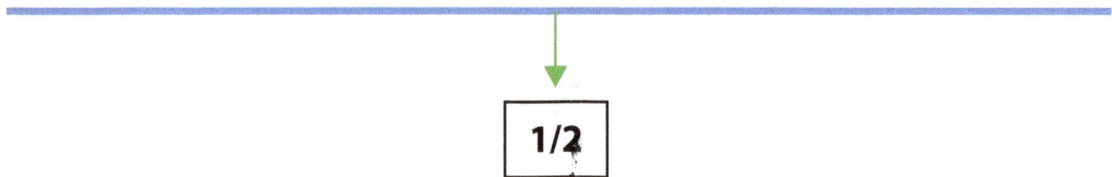

FIGURE 5-5 Length Model Example

Area Model

An area model is the most complicated form to represent fractions. Children in early and elementary levels would have some difficulty in understanding fractions using the area model. In an area model, you need to have the same area to represent fractions, but it is not necessary to have the same shapes. This type of model is not appropriate for young children who do not have an understanding of area. This would be more appropriate for upper elementary levels or children in advanced levels. As Figure 5-6 presents, each shape looks different but has the same area. In this figure, each shape represents ⅓.

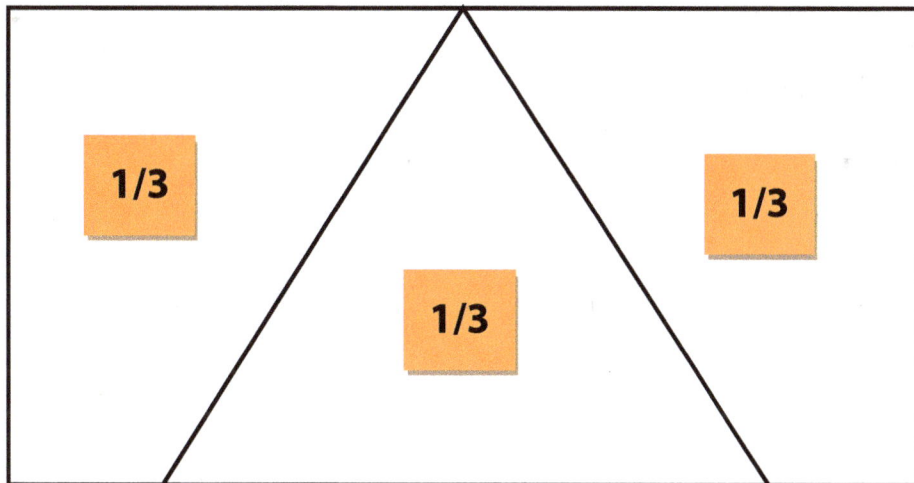

FIGURE 5-6 Area Model Example

Set Model

In a set model, a set of objects are considered as a whole. For example, Micaela has five candies, and she ate two candies. What fraction does it represent on how many candies Micaela ate? Set model is sometimes confusing for young children since they are still learning about whole number concepts. Children consider one candy as a whole instead of considering one as one-fifth of a set. It is necessary for teachers of young children to carefully plan the lesson if they are to utilize a set model to teach children fractions. The use of a set model is challenging to young children to understand the concept of fractions since a set model deals with counting objects (candies in Figure 5-8). When children count objects, they use whole numbers

Micaela had five candies.

She ate two candies.

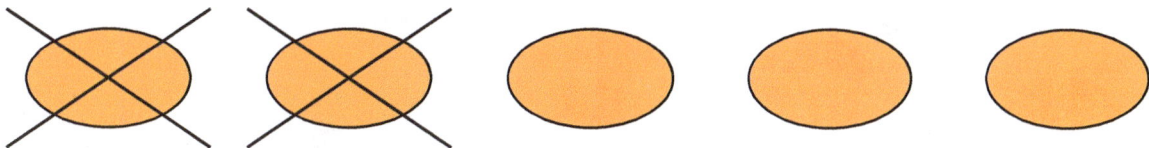

She ate 2/5 of candies. She has 3/5 of candies left.

FIGURE 5-7 Set Model Example

(e.g., 1, 2, 3, 4, and 5). Children would say she ate two candies out of five candies. Children understand the number of candies as whole numbers. However, as Figure 5-7 presents, Micaela ate two candies, which represents she ate 2/5 of the candies. This is confusing since children are still learning about whole numbers. In early childhood, it is recommended to apply a region model instead of using a set model.

Reflection Note:

Plan a math activity to develop children's number sense targeting first through second graders. See Section 5 for the components to consider in order to develop children's number sense.

Promoting Children's Algebraic Thinking and Teaching Algebra

At the end of Chapter 6, you should be able to:

- Describe how to promote children's algebraic thinking.

Chapter 6 presents how to promote children's algebraic reasoning and how to teach young children algebra. Table 6-1 presents NCTM's algebra standards and specific behavioral expectations for children pre-K through grade 2 and a Common Core Standards overview on "operations and algebraic thinking" from kindergarten through grade 2.

TABLE 6-1 Algebra Standards Grades Pre-K through 2

Algebra Standards	Behavioral Expectations for Pre-K through 2 (NCTM, 2000, p. 90)	Common Core Standards Overview on Operations and Algebraic Thinking (National Governors Association Center for Best Practice & Council of Chief State School Officers, 2010)
• Understand patterns, relations, and functions.	• Sort, classify, and order objects by size, number, and other properties; • Recognize, describe, and extend patterns such as sequences of sounds and shapes or simple numeric patterns and translate from one representation to another; • Analyze how both repeating and growing patterns are generated.	**Grade K (p.10)** • Understand addition as putting together and adding to, and understand subtraction as taking apart and taking from.

(Continued)

Algebra Standards	Behavioral Expectations for Pre-K through 2 (NCTM, 2000, p. 90)	Common Core Standards Overview on Operations and Algebraic Thinking (National Governors Association Center for Best Practice & Council of Chief State School Officers, 2010)
• Represent and analyze mathematical situations and structures using algebraic symbols.	• Illustrate general principles and properties of operations, such as commutativity, using specific numbers; • Use concrete, pictorial, and verbal representations to develop an understanding of invented and conventional symbolic notations.	**Grade 1 (p.14)** • Represent and solve problems involving addition and subtraction. • Understand and apply properties of operations and the relationship between addition and subtraction. • Add and subtract within 20. • Work with addition and subtraction equations. **Grade 2 (p.18)** • Represent and solve problems involving addition and subtraction • Add and subtract within 20. • Work with equal groups of objects to gain foundations for multiplications.
• Use mathematical models to represent and understand quantitative relationships.	• Model situations that involve the addition and subtraction of whole numbers, using objects, pictures, and symbols.	
• Analyze change in various contexts.	• Describe qualitative change, such as a student's growing taller; • Describe quantitative change, such as a student's growing two inches in one year.	

Algebra is often considered mathematics content for students in upper grade levels. The major reason for this is that many teachers of young children regard algebra as an abstract form of number operations and therefore as inappropriate for young children (Blanton and Kaput, 2003; Lee, et al., 2016). However, both NCTM (2000) and Common Core Math Standards (National Governors Association Center for Best Practices and Council of Chief State School Officers, 2010) include algebra in their math educational standards to be taught from early childhood.

Algebra reasoning evolves starting from pre-K and includes various mathematics activities such as sorting, classifying, ordering objects by a given attribute, patterns, basic operations, and so forth (Copley, 2010; NCTM, 2000). According to NCTM's position statement on algebra, all children should have access to algebraic concepts from pre-K years and appropriate supports for learning it from early years (NCTM, 2008). Common Core Math Standards present algebra to be taught by integrating number operations from kindergarten through grade 5 (National Governors Association Center for Best Practices and Council of Chief State School Officers, 2010).

Nevertheless, students in upper grade levels often express apprehension when they hear about algebra, in large part because of the way algebra is taught. Kilpatrick and his colleagues (2001) claim that this is because of the lack of coherence of teaching algebra and exposing children to it from early years. To provide a coherent and strong foundation of algebra, it is important to consider how to help children experience and learn algebra in a concrete manner and ultimately become fluent in algebra without losing their interest in their later school years. This chapter presents some practical ways to teach algebra to young children by promoting their algebraic reasoning.

6.1 Sorting and Classifying

The NCTM (2000) algebra standards for pre-K through grade 2 include the fundamental skills of sorting and classifying. Providing children various materials and having them to sort or classify them is a good activity to start with because it naturally enhances children's algebraic thinking. Sorting and classifying skills can be integrated into everyday routines without providing a structured mathematics lesson. For instance, a teacher can group children for collaborative work by the number of pockets they have in their jacket, by favorite color, favorite pets, and so forth. Another simple way to enhance children's sorting and classifying skills in daily routines is by labeling materials. For example, children should put their books in the labeled library corner in their classroom after reading. They also put any stationery (e.g., markers, papers, or scissors) in the right labeled spot.

There are several well-known mathematics manipulatives a teacher can integrate to systematically teach children classifying skills, such as pattern blocks or attribute blocks. Pattern blocks consist of different shapes and colors (e.g., yellow hexagon, orange parallelogram, etc.) that children use to make patterns as well as to classify by certain properties (e.g., things/shapes that fit with blue triangles). Attribute blocks are specifically designed to promote children's classifying skills and to promote their understanding of attributes of shapes. They consist of different colors, shapes, sizes, and thicknesses and allow children to practice classifying based on attributes. For example, a child may sort the attribute blocks based on one attribute and another child can guess at the target attributes. This can be extended to a fun game as children manipulate the materials.

6.2 Patterns

Patterns are the "fundamental concepts" of algebra and an important starting point from which to practice algebraic reasoning (Copley, 2000; McGarvey, 2012; NCTM, 2000; Taylor-Cox, 2003). Children experience patterns in their everyday life (e.g., shape patterns, number patterns, rhythm patterns, word patterns, etc.). Early childhood teachers often think they are not teaching "algebra" to their students even though they teach patterns every day. Patterns are extended to formulas in later grades. There are two types of patterns children can practice in early years using various concrete and abstract forms: basic repeating patterns and more advanced growing patterns.

Repeating Patterns

Children often identify patterns of color or shape. Pattern blocks are useful in teaching this method of identifying patterns (see Figures 6-1 and 6-2). The following figures present a repeating pattern. Having children repeat patterns using pattern blocks (or concrete materials) by color or shape helps them see and experience basic patterns and their relationships. Integrating music and song into mathematics classes is another fun way to teach about patterns. Body movement can also be easily integrated in mathematics lessons, such as by demonstrating a pattern of clapping and having the children follow the same pattern (e.g., clap, clap, tap, tap, clap, clap…). When the children line up at various times during the day, the teacher can ask them to stand in an AB pattern (e.g., girl, boy, girl, boy, etc.). As a group activity, a teacher can call a pattern and have a group of children make their own pattern by following the given patterns (e.g., AB pattern, AAB pattern, ABB pattern, etc.). For example, a teacher might call a pattern (AB pattern) to a small group of children. Children would come up with what attribute they would use as a group and create the pattern (e.g., first child raising right hand, second child raising left hand, third child raising right hand, etc.).

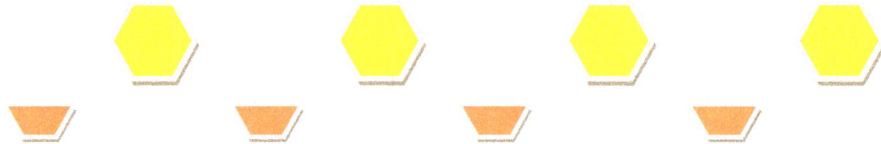

FIGURE 6-1 Repeating Pattern Type 1 (AB Pattern)

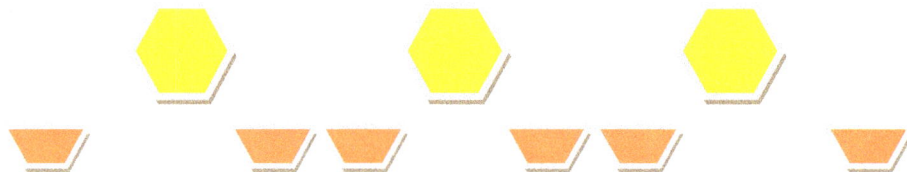

FIGURE 6-2 Repreating Pattern Type 2 (ABA Pattern)

Growing/Expanding Patterns

Once children become familiar with identifying repeating patterns, growing patterns can be introduced. While repeating patterns are a replicated form, growing patterns change or increase. For instance, as shown in Figure 6-3, the number of smiley faces is increasing. After showing children the figure below, a teacher may ask what comes next. To answer this question, children should identify the pattern first. Figure 6-3 has a growing number pattern of only smiley faces.

Figure 6-4 has a growing number pattern of two shapes: a smiley face and a star. When a teacher asks children to find this pattern, they need to know both the shape pattern and the number pattern.

FIGURE 6-3 Growing Pattern Type 1

FIGURE 6-4 Growing Pattern Type 2

In Figure 6-5, the number of mountains added on the bottom can be categorized as a growing pattern. The first picture has one mountain on the bottom, the second has two mountains on the bottom, the third has three mountains on the bottom, and the fourth has four mountains on the bottom. To find the fifth pattern, children should first identify the growing patterns. A teacher might also ask children to draw the pattern that would come in the sixth place to promote their algebraic thinking.

A teacher can also integrate number patterns (more abstract form of representation) without any pictures. With a 100 chart, children can easily find number patterns as they count by two or three. Daily calendar time is another good way for young children to practice patterns using days of the week.

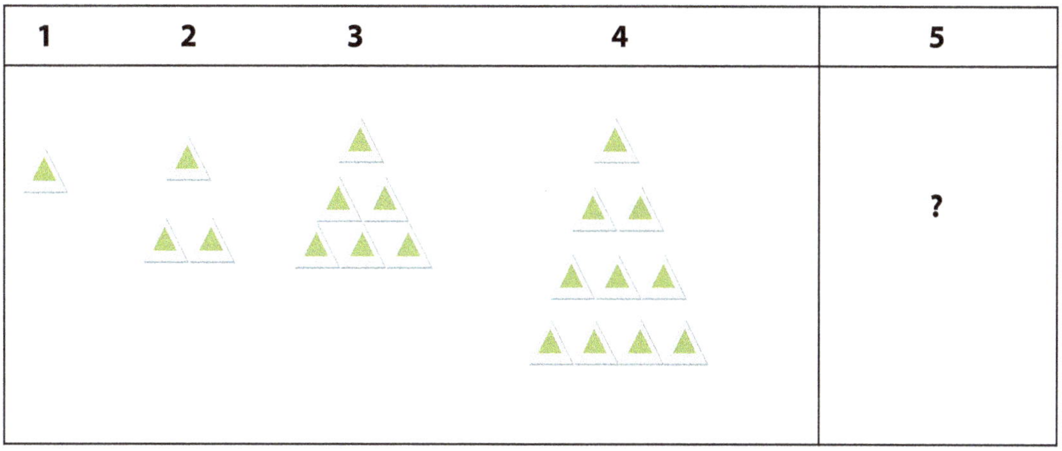

1	2	3	4	5
				?

FIGURE 6-5 Growing Pattern Type: Mountain Pattern, What's Next?

6.3 Understanding Functions: Relationships between Quantities

One purpose of algebra is to find the relationship between quantities. To help young children build knowledge and skills of understanding the relationship between quantities, it is necessary to show the relationship between quantities. One great activity to help children understand these relationships is to use a pan balance to measure the weight of two or more materials. Balancing activities have been found to be effective in helping young children to visually formalize the quantitative relationships between and among concrete objects (Curcio and Schwartz, 1997).

By manipulating a pan balance, children can reason the relationship among the shapes associated with quantities. For example, they can list heaviest to lightest weights. To make it more complicated, a teacher can add more shapes. To extend this concrete activity to more abstract thinking, a teacher of young children may use numbers to represent weight values, with a small number to be considered a lighter value and a big number to be considered a heavier value. This activity can be smoothly connected with algebra practice, enhancing children's concrete understanding of number operations and equality concepts.

6.4 Teaching Algebra to Children in the Upper Elementary Grades

TABLE 6-2 NCTM's Algebra Standards from Grades 3 through 5[1]

NCTM's Algebra Standards (NCTM, 2000, p. 394)	
Algebra Standards for Grades 3 through 5	**Behavioral Expectations**
• Understand patterns, relations, and functions.	• Describe, extend, and make generalizations about geometric and numeric patterns; • Represent and analyze patterns and functions using words, tables, and graphs.
• Represent and analyze mathematical situations and structures using algebraic symbols.	• Identify such properties as commutativity, associativity, and distributivity and use them to compute with whole numbers; • Represent the idea of a variable as an unknown quantity using a letter or a symbol; • Express mathematical relationships using equations.
• Use mathematical models to represent and understand quantitative relationships.	• Model problem situations with objects and use representations such as graphs, tables, and equations to draw conclusions.
• Analyze change in various contexts.	• Investigate how a change in one variable relates to a change in a selected variable; • Identify and describe situations with constant or varying rates of change and compare them.

Copyright © 2000, 2010 by National Council of Teachers of Mathematics (NCTM). Reprinted with permission.

Algebra for primary school grades integrates abstract forms of algebra involving simple operations, including symbols, functions, and equations. It is necessary for teachers of young children to help them practice finding the patterns and making generalizations based on the patterns they observed.

Numeric Patterns

In primary grades, assisting children to develop more comprehensive levels of algebraic thinking using more abstract forms—numbers and symbols—is critical:. Using a 100 chart (Figure 6-6), children practice finding number patterns by counting by a certain number. For example, ask children to count by four and mark each "4" on the

---

1   Copyright © 2000 by National Council of Teachers of Mathematics (NCTM). Reprinted with permission.

100 chart. They can easily see the pattern of counting by four and marking the numbers on the 100 chart. A teacher might also have children count by sixes or sevens and ask what changes they can see when counting by different numbers. In addition, encourage children to use a calculator to check their answers.

The next step to promote children's algebraic thinking is to gradually integrate more abstract forms such as tables, formulas, and so on. For example, have children determine the number of fingers in school. They can do this in a group. First, they need to fill out a chart like the one shown in Table 6-3 and find the number pattern to determine the total number of fingers in school. To make this task more challenging, you can ask children how many fingers N number of students have.

For example, let's say the number of students in school is 214. How many fingers are there in total in the school? Children need to find the patterns first and come up with their own formula, such as multiplying the number of students by 10 (number of fingers). Later, children need to come up with the number of fingers of unknown number (N) of students. Using the pattern, they should be able to come up with $10 \times N$ (or 10N).

1	2	3	4	5	6	7	8	9	10
11	12	13	14	15	16	17	18	19	20
21	22	23	24	25	26	27	28	29	30
31	32	33	34	35	36	37	38	39	40
41	42	43	44	45	46	47	48	49	50
51	52	53	54	55	56	57	58	59	60
61	62	63	64	65	66	67	68	69	70
71	72	73	74	75	76	77	78	79	80
81	82	83	84	85	86	87	88	89	90
91	92	93	94	95	96	97	98	99	100

FIGURE 6-6 100 Chart with Patterns

TABLE 6-3 Number of Fingers of One Hand in My School

Number of Children	1	2	3	4	5	......	214	N
Number of Hands	5	10	15	20	25	......	?	?

A more complicated problem can involve three-dimensional solids. For example, you can provide children cubes so they can explore the properties of the cube. As in Figure 6-7 below, children can find the pattern of the edges of cubes. First, they can count the number of edges of a cube, the number of edges of two cubes, and the number of edges of three cubes. They can keep working on finding the number of edges based on the number of cubes. Finally, children can find the number of edges of unknown (N) number of cubes represented by 12 × N (or 12N).

Cubes	Number of Cubes	Number of Edges
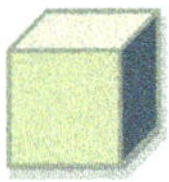	1	12
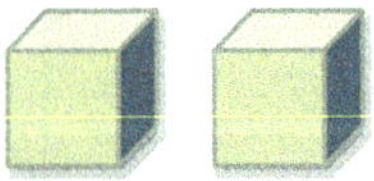	2	24
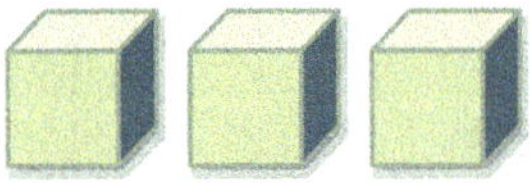	3	36
••••••	••••••	••••••
Unknown Number of Cubes	N	?

FIGURE 6-7 Number of Edges of Cubes

Exploring Equality

Provide various concrete experiences in order for children to explore relationships between objects. In the upper elementary grades, children should be able to use equations to demonstrate relationships. The most important concept used to explore relationships in algebra is "equality." A teacher will be able to provide concrete experiences of equality to children using a balance.

Analysis of Change

Children need concrete experiences to find numeral relationships. In grades 3 through 5, children are also able to analyze how one variable relates with another. The most effective way to show this principle is to help children become familiar with reading and understanding graphs (see NCTM, 2000, p. 163). Providing more concrete experiences to children is important before presenting abstract graphs. For example, you can use a plant that is growing in the classroom. Children can measure the height of the plant on a daily or weekly basis and write the data using a table. Later, they can report the growth of the plant using a form of graph to see how one variable (the number of days) relates with another (the height of the plant).

When introducing graphs, integrating real-life stories is an effective method to motivate children's interest in the target concept. Provided that you have the space, an interesting task that children always enjoy is Bungee Barbie or Ken. The major task is to find the appropriate number of rubber bands to maximize the thrill. This means that when children bungee the Barbie or Ken, the doll needs to come as close to the ground as possible without hitting the ground. Using rubber bands and different types of measuring tools (e.g., yard sticks, rulers, measuring tapes, etc.), the children can make a table to see the relationships (see Table 6-4). Table 6-4 shows the relationship between the number of rubber bands and the lengths of the bungee jumps Barbie or Ken make. You can also provide a height to children to bungee Barbie or Ken and have children estimate how many rubber bands they will need. Consider the place you are able to access with your children and have them actually bungee the Barbie or Ken to prove whether their plans work.

TABLE 6-4 Relationships between the Number of Rubber Bands and Lengths of Bungee Jumps

# of Rubber Bands	1	2	3	4	5	......	N
Length of Bungee Jumping	5 cm	11 cm	23 cm	47 cm	95 cm		500 cm

Once children have found the patterns between these two variables (the number of rubber bands and lengths of bungee jumps), they can draw a graph using the data they found. As the next step, they can predict how many rubber bands they would need to maximize Barbie's or Ken's thrills to get them close to the ground.

Reflection Note:

Observe a math lesson in an early childhood classroom (pre-K through grade 2 or 3) and reflect on how the teacher teaches algebra in the classroom.

CHAPTER 7

Teaching Children Geometry

TABLE 7-1 NCTM's and Common Core's Geometry Standards from Grades Pre-K through 12

Geometry Standards for pre-K through Grade 12	Behavioral Expectations for Pre-K through Grade 2 (NCTM, 2000, p. 96)	Common Core Math Standards Overview on Geometry (National Governors Association Center for Best Practice & Council of Chief State School Officers, 2010)
• Analyze characteristics and properties of two- and three-dimensional geometric shapes and develop mathematical arguments about geometric relationships	• Recognize, name, build, draw, compare, and sort two- and three-dimensional shapes • Describe attributes and parts of two- and three-dimensional shapes • Investigate and predict the results of putting together and taking apart two- and three-dimensional shapes	**Grade K (p.10)** • Identify and describe shapes. • Analyze, compare, and create, and compose shapes. **Grade 1 (p.14)** • Reason with shapes and their attributes. **Grade 2 (p.18)** • Reason with shapes and their attributes.
• Specify locations and describe spatial relationships using coordinate geometry and other representational systems	• Describe, name, and interpret relative positions in space and apply ideas about relative position • Describe, name, and interpret direction and distance in navigating spaces and apply ideas about direction and distance • Find and name locations with simple relationships such as "near to" and in coordinate systems such as maps	
• Apply transformations and use symmetry to analyze mathematical situations	• Recognize and apply slides, flips, and turns • Recognize and create shapes that have symmetry	
• Use visualization, spatial reasoning and geometric modeling to solve problems.	• Create mental images of geometric shapes using spatial memory and spatial visualization • Recognize and represent shapes from different perspectives • Related ideas in geometry to ideas in number and measurement • Recognize geometric shapes and structures in the environment and specify their location	

At the end of Chapter 7, you will be able to

- Describe geometric thinking for elementary children;
- Associate the van Hiele levels of thinking with how to teach children geometry;
- Describe how to increase children's geometric reasoning and spatial visualization skills using solid and plane geometry.

According to NCTM (2000), children should practice geometry beginning in their pre-kindergarten year. Common Core Math Standards also present specific geometry learning standards for all students from kindergarten through high school (National Governors Association Center for Best Practices and Council of Chief State School Officers, 2010). The above table presents the NCTM's and Common Core Standards on geometry for children from pre-K through grade 2.

The focus of geometry teaching practices in early childhood and elementary education has been placed on identifying formal attributes of geometric shapes rather than helping students develop a deeper understanding and reasoning of the properties and relationships of geometric objects by promoting their spatial sense. Before discussing how to teach children geometry aligned with standards, it is essential to consider how children think geometrically in order to be able to implement geometry lessons in a developmentally appropriate manner. This chapter focuses on Piaget's view of children's thinking associated with geometry and van Hiele's levels of geometric thinking.

7.1   Geometric Thinking

Piaget's View

Let's consider the ways children in pre-K through grade 2 think. Piaget, who was an advocate of cognitive development, believed that children ages two through six or seven are in a preoperational stage. In other words, they are unable to see a situation from another person's point of view. According to Piaget, the egocentric preoperational child assumes that other people see, hear, and feel exactly the same as he or she does. Piaget investigated this phenomenon in the so-called "mountains study." He put a child in front of a simple plaster model of a mountain range and seated himself to the side. Piaget then showed the child four pictures of the model taken from different points of view and instructed the child to "select the picture that shows what **I** see." Children in the preoperational stage selected the picture showing what **they themselves** saw, but older children selected the alternate viewpoint picture correctly. When teaching geometry to younger children, teachers should make sure to consider their thinking style and not assign tasks that require decentric thinking (defined as seeing an object from another's view). For example, showing one side of a geometric shape and asking about the properties of the other side of the object is too advanced for children in this stage because they cannot understand what is being asked.

In addition to egocentric thinking in the preoperational stage, teachers must also consider that children in this stage lack an understanding of the concept of conservation. They do not realize that when objects change in form, they do not necessarily change in amount. This lack results from not understanding the notion of reversibility, which is the understanding that certain processes can be undone or reversed. If you have two five-inch-long sticks parallel to each other, then move one of them a little (see Figure 7-1 below), a child in the preoperational stage may believe that the moved stick is now longer than the other. The child is unable to realize that this moving process can be reversed.

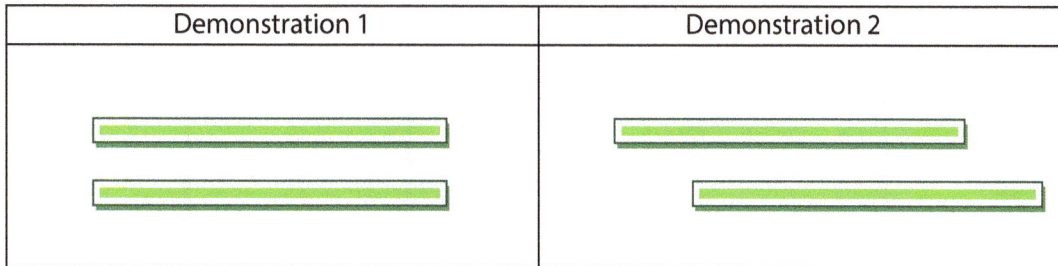

Demonstration 1	Demonstration 2

FIGURE 7-1 Piaget's Task-Length Conservation

Teachers of children in this stage should take this particular thinking process into consideration when teaching geometry. Young children have informal knowledge of geometry long before they come to school. They can differentiate between shapes and identify certain attributes (the number of points a shape has). Therefore, teachers of young children should help those children to bring and share their informal knowledge about geometry in class. For example, instead of presenting the concept of a triangle, you can ask children what they know about a triangle. Children might answer that it has three points. Elaborate the concept of triangle using the term "three points" instead of "vertex."

Van Hiele's View

Young children frequently perceive shapes holistically instead of seeing specific details. Children see a circle and may say, "It looks like a clock." Understanding how children think geometrically is an important first step to teach children geometry. Van Hiele presents five levels of children's understanding of geometry: visualization, analysis, informal deduction, deduction, and rigor. The first three levels are often integrated in early, elementary, and middle school levels. The following table presents an overview of van Hiele's levels of geometric thinking.

Shape	Point/Vertex	Side/Edge	Name	Other Properties
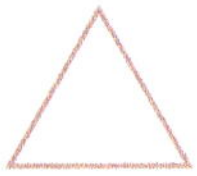 △				
 ▭				
○				
▢				

FIGURE 7-2 Similarities and Differences of Shapes

TABLE 7-2 Levels of van Hiele's Geometric Thinking

Levels	Description
Level 0: Visualization	See shapes as a whole without seeing properties of shapes.
Level 1: Analysis	Can identify the properties of shapes but are unable to see the relationships between shapes.
Level 2: Informal Deduction	Can identify the relationships between shapes.
Level 3: Deduction Level 4: Rigor	Children at both Deduction and Rigor levels are beyond the middle school levels. At the Deduction level, children can utilize deduction in an axiomatic system to prove statements (can solve proof-oriented problems taught at high school levels). Children at the Rigor level are able to see geometry in an abstract form.

At the visualization level, children need various experiences with shapes in their everyday lives. Since children at this level see a shape as a whole without being able to identify the properties of the shape, it is necessary to allow children to describe the shape in a holistic manner. For example, a teacher may ask children to find the rectangles in the classroom or on the playground. The teacher can explain the rectangle based on what children already know by asking what they know about it and go from there: "Yes, a rectangle looks like a table. Tell me about it. What do you know about the shape of a table?" The teacher can further children's understanding of the concept of rectangle by using questioning prompts. Children can also explore and search for shapes at home in everyday activities such as shopping for groceries. They can create an "I SPY SHAPES" book using the shapes they see in their lives. Children can either take photos or make drawings for their "I SPY SHAPES" book.

At the analysis level (approximately grades 2 to 3), children now can identify the properties of shapes, though they are unable to relate one shape to another. Teachers of children at this level can further children's understanding of properties of shapes using their own terms. For example, show a triangle and ask children to describe its characteristics. Children may answer by associating with quantity (e.g., it has three points). This is a teachable moment for teachers to introduce the school mathematics terms: "Yes, it has three points. We call each point a 'vertex.'" As always, start from where children are and build on it.

At the informal deduction level (approximately grades 3 and up), children are able to describe relationships between shapes. For example, children can tell that a square is a rectangle and are able to rationalize their thinking. Offer children at this level more opportunities to explore similarities and differences of shapes. For instance, children can easily point out the similarities and differences of given shapes. To make this learning experience more structured, provide a table with given shapes along with a list of properties for children to focus on (see Figure 7-2). This activity can bridge children's geometric thinking between levels 1 and 2.

Teachers need to provide geometry activities that help children progress from the level of thought at which they are operating to the next level. See the course page after this chapter for a link with more information on teaching practice based on van Hiele's model of geometric understanding.

Starting in grade 3, geometry content requires thinking and doing. Children in the upper grades will come to understand the relationships by doing activities involving sorting, building, drawing, modeling, tracing, measuring, and constructing. They also should be able to test, justify, and prove these relationships.

7.2 Geometry Standards Alignment

According to NCTM (2000), children should have experiences to analyze characteristics and properties of 2D and 3D geometric shapes, to explore and describe spatial relationships between geometric shapes, to use transformation, and to use visualization and geometric reasoning. Common Core Math Standards emphasize the importance of exploring shapes in early childhood by having children analyze, compare, create, and compose various shapes. It is also important for children to reason with shapes and their attributes (National Governors Association Center for Best Practices and Council of Chief State School Officers, 2010). Concrete experiences of manipulating, comparing, analyzing, composing, and decomposing various shapes build a strong foundation of geometric reasoning.

Transformation

Transformation is an important concept for children to understand in early and elementary levels. Transformation associated with hands-on activities enhances children's geometric reasoning as well as their spatial sense, which is defined as "awareness of children themselves in relation to the people and objects around them" (Copley, 2000, p. 105). Encouraging children to manipulate the various shapes helps them understand the concept of transformation. To build a strong conceptual understanding of transformation, first help them use their own physical experiences with shapes using their own terms or words such as slides (translations), turns (rotations), and flips (reflections).

One common activity in early childhood that promotes children's transformation practice is "shape puzzles," in which the child has to transform the shapes to complete the puzzle. As children move into grades 3 and up, they need to develop greater precision as they describe motions to show congruence (e.g., "turn it 90 degrees," "flip it vertically," then "rotate it 180 degrees"). They should also be able to describe motion using angles (e.g., "you need to turn 180 degrees around the center"). Children at these grade levels should be able to visualize and describe the relationship among lines of reflection, centers of rotation, and the position of pre-images and images.

Visualization

Although a teacher can help children learn geometric ideas by using physical materials, children still need to develop visual ideas of geometric shapes (e.g., making an open box using squares). As students mature, a teacher can encourage them to manipulate these images in their minds using questions such as "Which figure can be folded into the shape of the open box?" or "Which figure can be folded into a cube?" (see Figure 7-3). You can ask children to use their visualization skills to find answers.

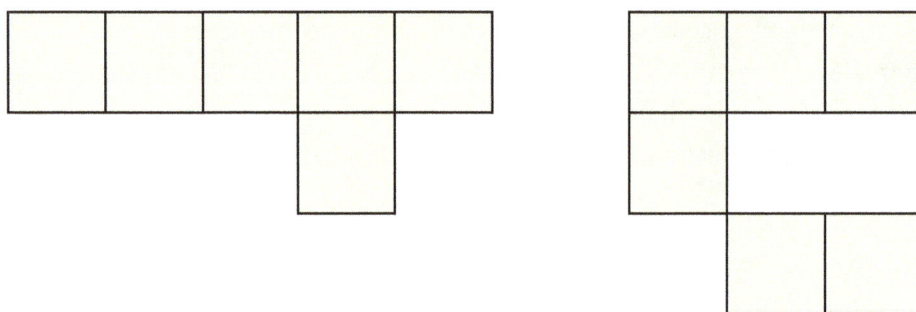

FIGURE 7-3 Visualization: Seeing Two Dimensional Nets and Visualization Open Boxes

Starting in grade 3, children need to practice visualizing three-dimensional to two-dimensional shapes and two-dimensional to three-dimensional shapes. You can ask children to make a building using ten cubes (see Figure 7-4). Children can manipulate three-dimensional shapes (solids) to make a building by looking at the two-dimensional shapes (plane shapes). For younger children, reduce the number of views and provide one view or two views to create the three-dimensional shapes (e.g., using only the top view or the top and front view).

Top View

Front View

Right Side of View

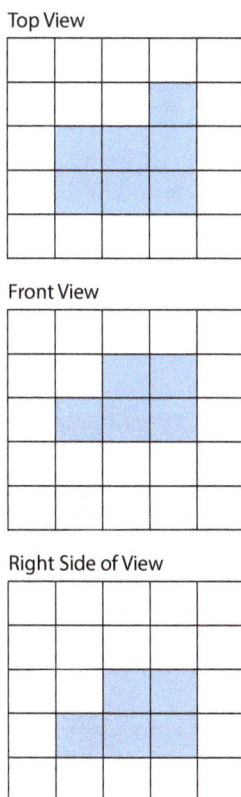

FIGURE 7-4 Visualization: Two-to Three-Dimensional Shapes

Copyright © 2000, 2010 by National Council of Teachers of Mathematics (NCTM). Reprinted with permission.

7.3   Geometry Manipulatives

Pattern Blocks

One of the most commonly used materials in elementary school is pattern blocks. There are usually six basic shapes in a set of pattern blocks, such as square, triangle, two types of rhombuses, hexagon, and trapezoid. Pattern blocks can be used several ways. One is to let children play with the blocks to design or create artwork. This promotes children's spatial sense and awareness of the attributes of shapes. By adding matching cards in pattern block play, children can practice observing properties of pattern shapes to match the cards.

Geoboard

A geoboard is a manipulative that helps children explore basic concepts in plane geometry such as perimeter, area, or the characteristics of shapes. A geoboard consists of a physical board with a certain number of pegs and rubber bands. Encourage children to create any shapes using rubber bands on the geoboard and have them share what they created. Since they are using rubber bands, it is necessary to share safety rules (e.g., do not make a shape too big with one rubber band, do not use it as a play material but use it to make a shape on the geoboard, etc.). A geoboard is well-designed to show children a vertex model. For example, the teacher can ask students to make a triangle. To make a triangle, children have to consider how many points (vertexes) a triangle has and then use a rubber band to make a triangle by placing a rubber band on three points.

A little more complicated problem will promote children's geometric problem-solving skills. Ask students to make a square that contains four pegs. To make a square, children have to consider how many edges a square has, then use a rubber band to make a square. Geoboard activities help children become familiar with the names of shapes as well as properties such as edges and vertexes. Figure 7-5 presents a picture of a geoboard with rubber bands.

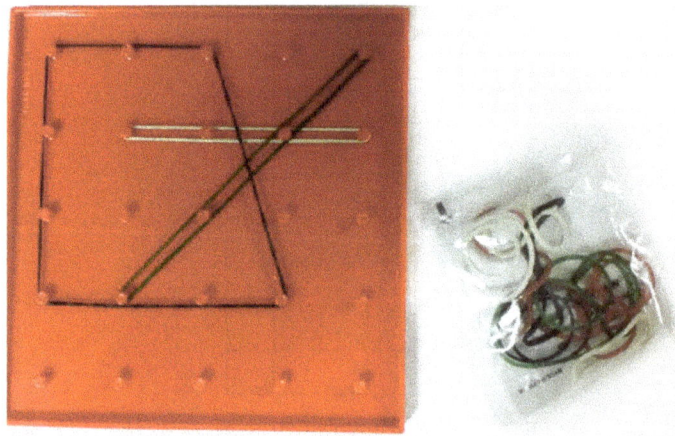

FIGURE 7-5 Making a Triangle with a Geoboard

Attribute Blocks

Attribute blocks are a great manipulative that helps children become familiar with the properties and attributes of shapes as they explore and play with the blocks. Attribute blocks help children practice sorting blocks by color, shape, and size. Basic sets of attribute blocks have sixty pieces in five shapes (circles, squares, hexagons, rectangles, and triangles), three colors (usually yellow, blue, and red), and two thicknesses (thin and thick) that can be sorted in several ways. Pull out some attribute blocks and display them on the floor for children to see. Tell the children that you will sort the blocks in a certain way, then show children blocks sorted into two sets. Finally, ask children to figure out what common attribute/characteristic is used. Children use this activity as a geometry puzzle. A child might sort the attribute blocks using one attribute without telling it. Other children may find the attribute the child used for sorting. This type of activity promotes children's geometric thinking and geometric reasoning skills.

FIGURE 7-6 Attribute Blocks. You can check out an "Attribute Block Interactive Online Activity" at http://nlvm.usu.edu/en/nav/frames_asid_270_g_2_t_3.html?open=instructions

Mira/Geo-Reflector

Finding the symmetry line of shapes is sometimes challenging for young children. To help children build a conceptual understanding of symmetry lines, it is necessary to provide concrete experiences. For example, have children practice a "fold and match" activity by providing various shapes and having them match the shapes by folding them in half. A decalcomania activity helps children see the symmetry lines of various shapes as well. A "mira" or "geo-reflector" is a specific tool to help children practice drawing symmetry lines. By using the geo-reflector, children can easily draw congruent shapes by looking at the reflected shape on the mira/geo-reflector and copying it. In addition, the mira/geo-reflector helps children find the symmetry line by moving it to match the reflection of the shape.

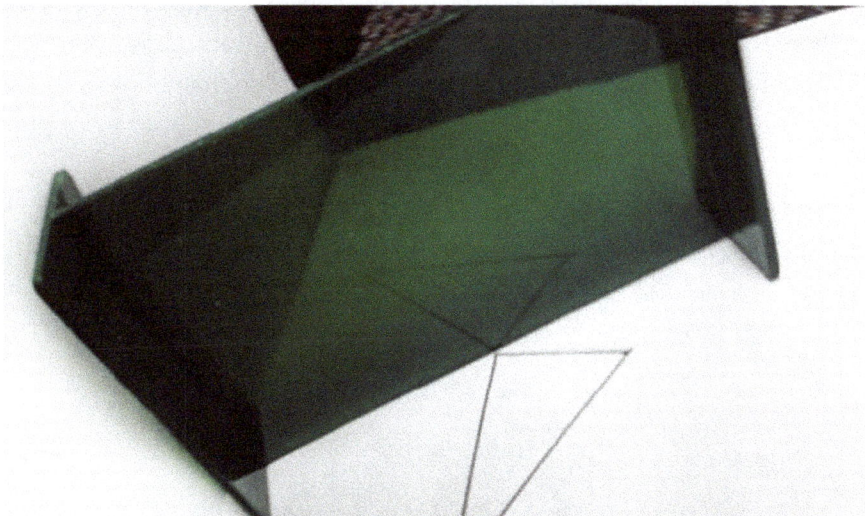

FIGURE 7-7 Mira/Geo-Reflector

Tangram

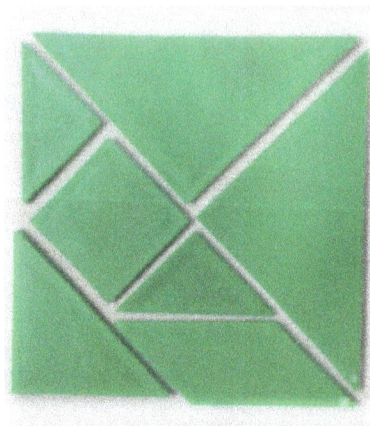

FIGURE 7-8 Tangram

A tangram is a very old Chinese puzzle consisting of seven pieces that can be arranged to make many different pictures. A tangram set is made from one square cut into seven pieces: two big triangles, two small triangles, one parallelogram, a square, and a medium triangle. There are many ways to use tangrams to promote children's spatial sense. Since this is an open-ended material that has various uses and purposes, tangrams have been utilized from early childhood to college level (Lee, et al., 2009). A tangram is specially designed to enhance spatial sense by understanding the relationship between and among shapes. Providing tangrams during center time helps children explore, manipulate, and create tangram shapes based on their interests without any directions. To extend tangram activities, first have children create various shapes or copy shapes of their friends' creations. Having children create a certain shape helps them practice transformation of shapes (e.g., creating a square with two shapes, three shapes, etc.). Integrating children's books easily motivates children's interest in tangrams even more. There are many children's books associated with tangram illustrations (e.g., *Tang's Story* by Ann Tompert; *Tangram Magician* by Lisa Campbell Ernst; *Three Pigs, One World, Seven Magic Shapes* by Grace Maccarone and David Neuhaus). While reading the book, the children can make the shapes shown on each page. Later on, have children create their own story or poem using tangram illustrations and publish their books to put in a classroom math library.

Reflection Note:
Reflect on how van Hiele's levels of thinking are associated with teaching children geometry.

Measurement for Young Children

At the end of Chapter 8, you should be able to:

- Differentiate between standard and non-standard units of measurement;
- Explain how to promote children's understanding of needs of standard units of measurement.
- Explain how to promote children's measurement reasoning.

NCTM presents measurement standards for children from pre-K through grade 12 (see Table 8-1 for standards along with behavioral expectations for children from pre-K through grade 3). In addition, it also presents the Common Core Standards overview on "measurement and data" from kindergarten through grade 2.

TABLE 8-1 NCTM's & Common Core's Measurement Standards

Measurement Standards for pre-K through Grade 12	Behavioral Expectations for Pre-K through Grade2 (NCTM, 2000, p. 102)	Common Core Standards Overview on Measurement and Data (National Governors Association Center for Best Practice & Council of Chief State School Officers, 2010)
• Understand measurable attributes of objects and the units, systems, and processes of measurement.	• Recognize the attributes of length, volume, weight, area, and time; • Compare and order objects according to these attributes; • Understand how to measure using nonstandard and standard units;	**Grade K (p.10)** • Describe and compare measurable attributes. • Classify objects and count the number of objects in categories. **Grade 1 (p.14)** • Measure lengths indirectly and by iterating length of units. Tell and write time • Represent and interpret data.

(Continued)

Measurement Standards for pre-K through Grade 12	Behavioral Expectations for Pre-K through Grade2 (NCTM, 2000, p. 102)	Common Core Standards Overview on Measurement and Data (National Governors Association Center for Best Practice & Council of Chief State School Officers, 2010)
	• Select an appropriate unit and tool for the attributes being measured.	Grade 2 (p.18). • Measure and estimate lengths in standards units. • Relate addition and subtraction to length. • Work with time and money. • Represent and interpret data.

Copyright © 2000, 2010 by National Council of Teachers of Mathematics (NCTM). Reprinted with permission.

In teaching children measurement, building a connection between non-standard and standard units of measurement is important. Children come to school with some level of confidence and familiarity of measuring things using non-standard units of measurement such as length, area, weight, and time. The following session presents how to promote children's measurement reasoning by connecting from non-standard to standard units of measurement.

8.1 Using a Non-Standard/Arbitrary Unit

Before children use a standard unit of measurement, they often use a non-standard/arbitrary unit to measure attributes of things they see in their daily lives. For example, children often compare their height with their friends; who is taller is sometimes a major concern among children. Though children don't use a standard unit of measurement such as ruler or measuring tape, they are still able to measure the heights of themselves and others. Children often use a reference to measure. Using each other's height ("reference"), children decide who is taller or shorter. Children are often interested in measuring which of two items is longer or shorter. The same principle applies for measuring other attributes such as weights (heavier or ligher) or areas (bigger or smaller). These everyday experiences should be brought to early childhood classrooms to extend children's understanding of non-standard unit measurement. Integrating daily life measurement experiences into math lessons often makes math learning meaningful for children. It is necessary to provide various concrete items for children to manipulate to practice measurement, including length, weight, volume, capacity, and time.

8.2 Measuring Attributes in Early Childhood

In order for children to understand measuring, they first have to identify what to measure, which is called "measuring attributes." In other words, children must identify attributes to be measured before they start to measure. Young children from pre-K through grade 2 can compare attributes by looking at physical properties. According to Common Core Math Standards, it is important for young children to sort and to classify objects by "measureable" attributes, which promotes children's measurement reasoning. For example, you can ask children to compare two objects based on length, thickness, or some other attribute. Children in these grade levels can compare the attributes using terms like "shorter," "taller," "longer," "thicker," "thinner," and so on. Expose children to various objects such as pencils, crayons, spaghetti, or Cuisenaire rods (a type of mathematics manipulative composed of different sizes of rods) so they can explore various measurable attributes by practicing measurement.

Length

Length is the easiest attribute for young children to practice measuring since it is one-dimensional (Copley, 2010). When they are first learning to measure length, children often use non-standard units such as paper clips, pencils, or shoes to measure a given item. Provide children in pre-K through grade 2 with different types of non-standard/arbitrary units. For example, ask children to measure the length of an item (e.g., table, door, blackboard, etc.) by selecting appropriate arbitrary units (paper clips, pencils, shoes, etc.). As children grow, they gradually begin to practice using standard units of measurement from second grade. When children practice measurement using non-standard units, they follow certain steps (see below). These steps are the exactly the same steps children take when they measure an attribute using a standard unit. Practicing measurement by following the steps builds a concrete foundation for children when measuring attributes using standard units. The following is a list of the steps children need to know and practice when they measure using either a non-standard/arbitrary unit or standard unit.

- Step 1: Know what to measure/Define the attribute to measure: Children need to identify what to measure (attributes), such as length, weight, capacity, area, volume, time, etc.
- Step 2: Know what unit to use: Children need to decide on an appropriate unit to measure the target attribute.
- Step 3: Know how to measure: Children need to know how to use the measuring tool.
- Step 4: Record and report the finding: Children need to record their measuring result(s).

Capacity

Capacity is defined as "the amount that something can hold." Capacity is not an easy concept for young children to grasp unless it is introduced in a developmentally appropriate manner. Capacity is an attribute that children often enjoy measuring from early years, though many teachers of young children might not be aware which activities demonstrate children measuring capacity. In a pre-kindergarten classroom, it is a frequent observation that children use various sizes of containers to scoop sands, macaroni, or water in a sand table or water table. This is also observed during outdoor play. These types of activities help children build concrete understanding of capacity and help them practice measurement of capacity using non-standard units. Some questions to promote children's measurement reasoning of capacity include: How many scoops of sand do you need? Which scoop would hold more sand (or water)? Encourage children to use various sizes of containers and compare the capacity of those containers. Also, providing children marbles/beads and cups with different sizes (small, medium, and large) to compare the capacity of each cup.

Young children use perceptual comparisons between two containers. Sometimes children have difficulty in comparing two different containers because they tend to see only one dimension, particularly height. Consider Piaget's volume conservation task. The same rule applies to capacity conservation. For example, put the same number of beads into a bottle that is long and narrow and a bottle that is short and wide. If you ask very young children which bottle holds more, they might say the one that is long and narrow. Providing various types of containers with dry materials or liquids allows children to explore capacity in a concrete manner.

Weight

In early and elementary levels, a teacher needs to provide various opportunities to measure weights. Children enjoy measuring weights using non-standard units. Children often use their hands as a balance to compare weights of different objects and should be able to identify which is heavier and lighter. When children are unable to determine the differences, this is a teachable moment by providing children a pan balance to compare the weights of both objects. Making a connection with children themselves is one effective way to promote children's interests in measurement and to promote their measurement reasoning (e.g., measuring children's own weights). Common Core Math Standards present the need for integration of measurement with other math content such as data representation and interpretations—by the second grade, children practice representing and interpreting data with measurement data. For example, once children measured their weights, collect the data of all children's weights, and represent/interpret the data of class weights.

Area

Area is defined as "the size of a surface," and it frequently uses square units to measure. Children can perceptually compare the area between two plane shapes by using terms like "larger" or "smaller." Without any formula, children still practice measurement of area. In early childhood, it is recommended to use actual squares (inch squares or other sizes of squares) to build conceptual understanding of area. Have children use squares to cover up a surface of a shape to measure its area (e.g., how many squares do you need to cover up this shape?/I need four squares to cover this up—this means that the area of the shape is four square units). This can be considered as a counting activity using the number of squares, but it helps children build understanding of area.

Volume

The amount of three-dimensional space an object takes up is its volume. Because "volume" is an abstract term, this concept is generally introduced in grade 4 or 5. However, in the early years, children need to explore and play with different shapes and sizes of containers filled with various sizes of cubes. Since volume is measured by cubic units, it is recommended to help children practice measuring volume using cubes.

Temperature

Temperature is the indication of how hot or cold a thing is. Temperature should be introduced to young children by integrating knowledge from their daily lives. Children often tell how cold or how hot something is. Integrating children's daily words and bringing their experiences to the class is important to help children understand the concept of temperature. Though children might not understand what it means to read a thermometer to measure temperature, continue to expose children to a thermometer to measure the temperature every morning and record it daily. This helps children gradually understand the relationship between numbers from a thermometer and temperatures.

Time

According to Common Core Math Standards, children from first grade should learn to tell and write time. Children are exposed to time in their everyday life. However, they often encounter challenges when it comes to telling and writing time in a formal manner. Time is a challenging concept for young children since it involves measurement units. Time is everywhere children go, but it is such an abstract concept. Children have difficulty measuring time since it includes two or more indicators of measurement, each of which uses different measurement units (e.g., second, minute, and hour). In addition, these indicators move in a circular manner, which even further challenges children since they are familiar with linear measurement (Reys et al., 2002). Piaget (1969) recommends helping children understand the concept of time by the sequence of events. Caplan and Caplan (1983) also suggest using terms based on sequential events (e.g., using the terms "first," "second," etc.) in early childhood before children are introduced to telling and writing the time. Also, connecting the events to certain times is suggested. In addition, Lee and her colleagues present effective ways to help children learn about time: sharing the daily schedule with children based on the sequence of events and time; displaying clocks (both analog and digital) at children's eye level; and having calendar time (Lee et al., 2009). One popular activity in early childhood classrooms is calendar time, which can easily be used to expose children to the concepts of date, month, number of days in a month, and so on. Caplan and Caplan (1983) suggest that children understand time concepts based on sequential events (e.g., using the terms "first," "second," etc.) before they understand the length of time and connecting events to certain times. Use sequential words to describe events and time.

8.3 Connecting Arbitrary Unit (Non-Standard Unit) with Standard Unit

In early childhood, helping children to bridge the ideas of a non-standard unit and a standard unit is essential. *How Big Is a Foot?* by Rolf Myller is a great children's book to help children understand both non-standard and standard measurement units and the needs of standard units of measurement. This book facilitates rich discussions that can help children to think about the challenges of using arbitrary units and the need for standard units. In the story, a king orders a carpenter to make a bed for his queen. The king uses his foot as a reference and tells the carpenter the size of the bed. However, the carpenter uses his own foot to make the bed. When the bed is delivered, it is too small for the queen. This is the challenge presented by the book. You can open discussion about what happened with this bed.

Before reading the book to children, ask them to work out the same mathematics problem as in the story. For example, "I need a bed for my mom and I'd like you to make the bed for her. Here is what she wants. She said that she needs a bed that is 9 feet by 7 feet." The children will most likely not understand the meaning of "foot." This task would open various discussions such as whose foot they need to use, whether they can use a measuring tool, and so on. This is a task that helps children see why they need a standard unit to measure. As a teacher, you need to elaborate the concepts of non-standard and standard units and explain why they need to know and use a standard unit to measure.

FIGURE 8-1 Child with a Measuring Tape

Copyright © Depositphotos/Demixx.

8.4 Standard Units

Children in the primary levels should be able to identify the standard units of measurement. According to Common Core Math Standards, children in second grade should be able to "measure and estimate lengths in standard units" (National Governors Association Center for Best Practices and Council of Chief State School Officers, 2010, p.18). A standard unit of measurement should include a number and measurement unit.

Familiarity with using units in measuring will help them later. For example, children might say only the number to measure the area (e.g., "the area of this rectangle is eight"), but this does not provide any information. They must learn that they need to include the unit. Emphasize that children need to use both number and unit in measuring. In earlier grades, get children in the habit of using whatever arbitrary/non-standard unit they are using to measure, such as paper clips or pencils (e.g., "nine paper clips long"). This practice will eventually help children get in the habit of using both numbers and units when they begin using standard measurements.

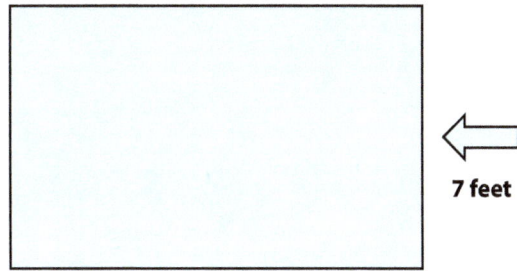

FIGURE 8-2 Measuring a Bed

Reflection Note:

Look for information on when to teach metric units and customary units of attributes (length, weight, capacity, area, volume, temperature, and time). Identify grade levels or ranges for which metrics and units should be taught. See the following table (Table 8-2) and fill out the blank columns. The NCTM standards, Common Core Math Standards, state standards, and your school district curriculum would be good resources to look for information. This activity allows you to become familiar with standards and measurement standards.

TABLE 8-2 Standard Units for Elementary Education

Attribute	Metric Units	Grades	Customary Units	Grades
Length	Decimeter Centimeter Meter Millimeter		Inch Foot Yard Mile	
Weight	Kilogram Gram		Ounce Pound	
Capacity	Liter Milliliter		Quart Cup Gallon	
Area	Square Centimeter Square meter		Square inch Square foot Square yard	
Volume	Cubic centimeter Cubic meter		Cubic inch Cubic foot Cubic yard	
Temperature	Celsius degree		Fahrenheit	
Time			Hour Minute Second Day Week Month	

Adapted from: Van de Walle, J.A. (2004). *Elementary and middle school mathematics: Teaching developmentally.* Boston: Pearson Education, Inc.

Data Analysis and Probability for Children

At the end of Chapter 9, you should be able to:

- Describe how to promote children's data analysis skills;

- Pose an important question for children to further their skills of data analysis;

- Explain how to promote young children's (pre-K through grade 2) understanding of probability.

Table 9-1 presents NCTM's data analysis and probability standards, behavioral expectations for children from pre-K through grade 2, and an overview of Common Core Math Standards on measurement and data.

TABLE 9-1 Data Analysis and Probability Standards

NCTM's Data Analysis and Probability Standards for Grades Pre-K through 12	Behavioral Expectations for Pre-K through 2 (NCTM, 2000, p.108)	Common Core Overview of Measurement and Data Standards
• Formulate questions that can be addressed with data and collect, organize, and display relevant data to answer them • Select and use appropriate statistical methods to analyze data	• Pose questions and gather data about themselves and their surroundings • Sort and classify objects according to their attributes and organize data about the objects • Represent data using concrete objects, pictures, and graphs • Describe parts of the data and the set of data as a whole to determine what the data show;	Grade K • Describe and compare measurable attributes • Classify objects and count the number of objects in categories Grade 1 • Measure lengths indirectly and by iterating length units • Tell and write time • Represent and interpret data

(Continued)

NCTM's Data Analysis and Probability Standards for Grades Pre-K through 12	Behavioral Expectations for Pre-K through 2 (NCTM, 2000, p.108)	Common Core Overview of Measurement and Data Standards
• Develop and evaluate inferences and predictions that are based on data • Understand and apply basic concepts of probability	• Discuss events related to students' experiences as likely or unlikely	Grade 2 • Measure and estimate lengths in standard units • Relate addition and subtraction to length • Work with time and money • Represent and interpret data

The content for data analysis and probability is often considered appropriate for upper elementary, middle, or high school levels. However, NCTM addresses the need to teach these contents starting from pre-K (see NCTM's data analysis and probability standards), and Common Core Math Standards on data highlights the need for children to practice collecting, representing, and interpreting data from early years (see Table 9-1).

Probability involves other mathematics (e.g., measurement) and computation skills such as counting, adding, subtracting, and averaging, and data analysis, statistics, and probability are often connected with other mathematics topics as well as other school subjects that provide wonderful learning and teaching opportunities for curriculum integration.

9.1   Formulating Questions and Answers Using the Data

To help children collect data in a meaningful way, it is necessary to associate the data with children's surroundings and interests. For instance, students can collect information about their families or friends and discuss the data in class. The class as a whole can make a concrete graph or pictograph to visually demonstrate the data. This practice helps children see the process of collecting, representing, and interpreting data in a meaningful manner. It is also important for teachers to encourage children to come up with important and measurable questions that can be answered by collecting data. Have children practice posing questions that interest them and that are not obvious and facilitate discussion on how to find the answers.

Once they have come up with a good question, they need to discuss what data to collect to answer the question and how to collect the data. This discussion allows children to engage in mathematics communication, which is strongly recommended by the NCTM and Common Core Standards. Here is one example that promotes children's mathematics communication on data. Kindergarteners may want to have a class pet. Before they adopt the class pet, they need to decide what type of animal to adopt. This is a good question that motivates children's interest and is not obvious to answer without data. Facilitating class discussions helps children come up with ways to solve the problem by considering various factors of animal life (e.g., food, habitat, life cycle, etc.). Once children shorten the list of animals, they now need to decide which animal to pick for the class pet. Based on this discussion and voting process, children can easily practice collecting, representing, and interpreting the data. When representing the data, using either a real/concrete graph or pictograph helps children read and interpret the data concretely. Creating a picto-table is another activity that can help children see that data concretely and also build a foundation for later learning about data analysis. A picto-table is a table using pictures, which is easier for young children to understand. Children are easily able to use a picto-table in voting by placing a Popsicle stick in the column for their favorite pet (see Figure 9-1). Having children put Popsicle sticks in the column of their favorite pets allows them to visualize the data and tell which pet is the class favorite. Using a picto-table or concrete data table provides children fundamental skills and understanding of an abstract form of data table they will see in their later school years.

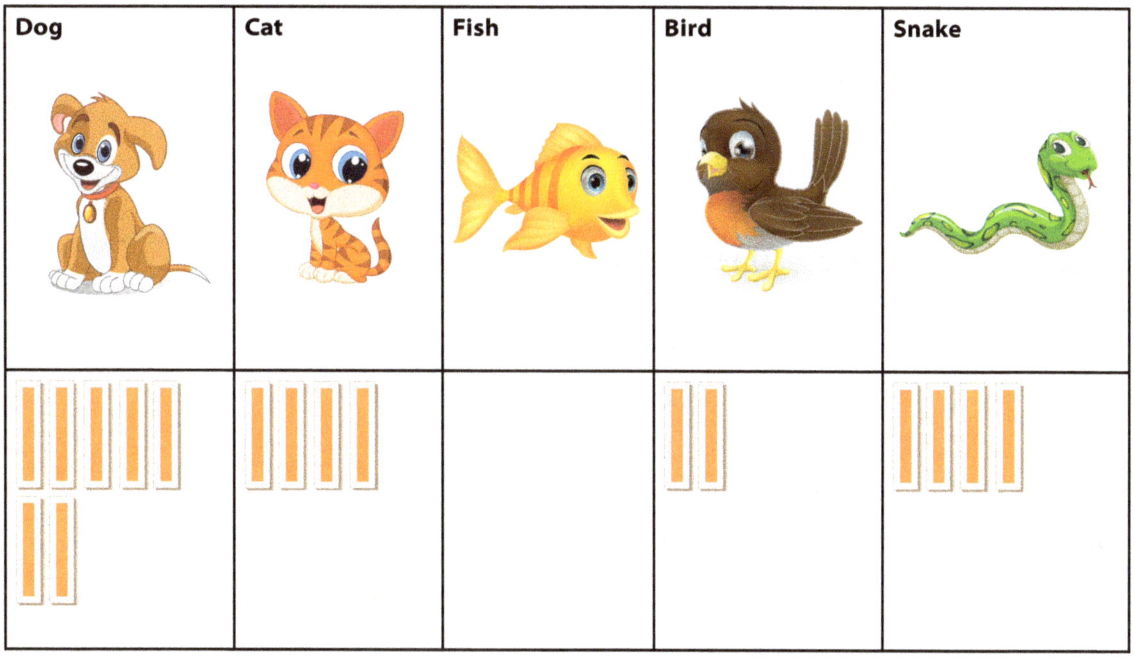

FIGURE 9-1 Class Favorite Pet Data—Dog, Class Favorite Pet Data—Cat, Class Favorite Pet Data—Fish, Class Favorite Pet Data—Bird, Class Favorite Pet Data—Snake

Credits: Dog: Copyright © Depositphotos/Dazdraperma.
Cat: Copyright © Depositphotos/tigatelu.
Fish: Copyright © Depositphotos/tigatelu.
Bird: Copyright © Depositphotos/Krisdog.
Snake: Copyright © Depositphotos/interactimages

9.2 Selecting and Using Appropriate Statistical Methods to Analyze Data

In pre-K through grade 2, children are interested in data acquired from their own lives. The data often used with these children are individual or discrete pieces of data or "value with the most" (favorite pets, transportation they use to come to school, favorite color, etc.). In grades 3 through upper elementary and middle levels, children should be able to compare the data sets, noting the similarities and differences between them. Children in these grade levels become familiar with precisely describing the data sets and are able to use major central tendencies: mode, median, and mean.

9.3   Displaying Data

Once children collect the data, they should be able to present the data using various types of graphs:

- Real graph: easy for young children to use and interpret. Children use actual objects to create the graph.
- Pictograph/picture graph: easy for young children to use and interpret. Children use pictures to create a graph.
- Bar graph: used for discrete data (e.g., favorite pets, favorite foods, favorite seasons, etc.). Children can easily interpret data by comparing the length of each bar.

Line Plot of Number of Pockets in Individual Child's Jacket

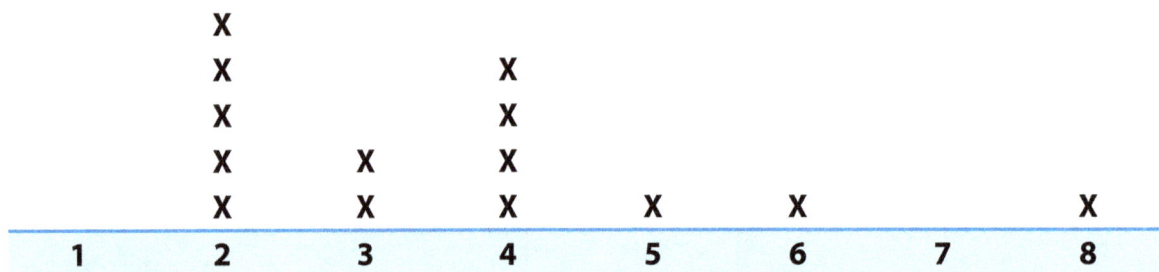

```
    X
    X           X
    X           X
    X     X     X
    X     X     X     X     X           X
----------------------------------------------
 1     2     3     4     5     6     7     8
```

FIGURE 9-2 Line Plot of Number of Pockets in Individual Child's Jacket

- Line plots and Stem-and-Leaf plots: a convenient method for displaying numerical data with a small range. This graph is appropriate for primary grades and up (see Figure 9-2).
- Pie graph: a form of circle graph. The circle represents the whole (100%), so when using a pie graph, you need to have exactly 100%. The pie graph is easy to understand and shows fractional parts based on a whole or 100%. A pie graph is appropriate for primary grades and up (see Figure 9-3)
- Line graph: used to show data trends over a period of time. Continuous data is appropriate for a line graph. A line graph is more appropriate for primary grades and up.
- Box plot: also called a "box-and-whisker plot." The box plot effectively shows how data are spread over time. In a box plot, a median is the important reference to compare. A box plot requires a certain level of statistical knowledge of range, median, and quartiles to interpret the data. For example, height, weight, or the scores of standardized tests/achievement tests would be good data to display with a box plot. Check out the following link to explore and manipulate box plots: http://nlvm.usu.edu/en/nav/frames_asid_200_g_4_t_5.html?open=instructions
- This graph is more appropriate for children from grade 6 and up since it involves some level of statistics.

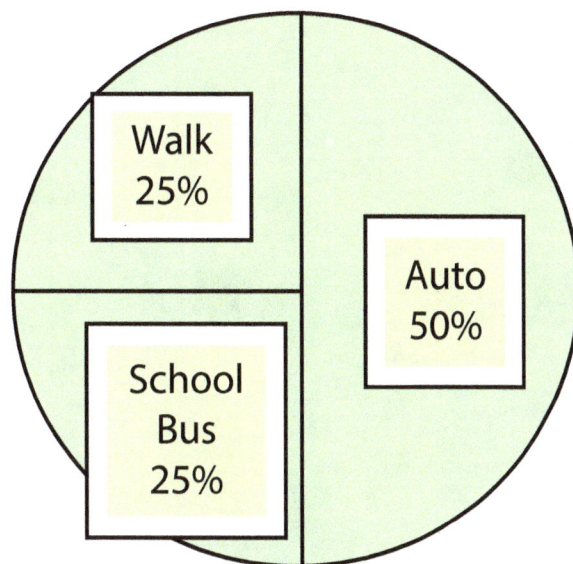

FIGURE 9-3 Pie Graph: How to Come to School

9.4 Teaching Children Inference and Probability

To teach young children probability, the NCTM recommends exposing them to events associated with experiences based on likelihood, such as likely, more likely, less likely, or unlikely. Children can use inference skills to answer the likelihood (e.g., likelihood of snow falling in Texas or Florida). This process promotes children's probabilistic reasoning. Inference and prediction are concepts requiring more advanced skills in mathematics, but concrete experiences in early years help children to build the foundational skills of inference and prediction that get them to an advanced level of data analysis and probability skills. Probability is based on data. As children concretely experience collecting, representing, and interpreting data from their early years, these practices build a strong foundation of later knowledge of data analysis and probability. In early grades, spinners or dice can be used effectively to help children understand probability. As children spin a spinner or toss dice, encourage them to record the results, which is the basic concept of "experimental probability" (vs. theoretical probability), and discuss the results in class. It is critically important for young children to practice probability from early years to be able to build a strong foundation of probability.

Reflection Note:

Observe an early childhood classroom and reflect on how the teacher implements a lesson on data analysis and/or probability.

Assessing Children's Mathematics Learning

At the end of Chapter 10, you should be able to:

• Differentiate between formative and summative assessment;

• Identify effective assessment strategies to assess children doing mathematics.

Assessment is the "process of gathering evidence about a student's knowledge of, ability to use, and disposition toward mathematics and of making inferences from that evidence for a variety of purposes" (NCTM, 2000, p. 3). Assessment cannot be separated from instruction but should be considered an integral part of instruction that benefits both teachers and students. According to the NCTM, "Assessment should support the learning of important mathematics and furnish useful information to both teacher and students" (NCTM, 2000, p. 22). The NCTM suggests that assessment should have two major principles: it should enhance students' learning, and it is a valuable tool for making instructional decisions.

Assessment should evaluate four major areas: student learning, student progress, effectiveness of instruction, and effectiveness of program (Mindes, 2003). It is important for teachers to continuously assess children in various ways. All assessments are categorized as one of two types: informal (formative) and formal (summative). Informal assessment takes place as teachers constantly evaluate children's learning by continuously checking and monitoring children's work and paying close attention to student reactions or responses to the teacher's questions during the lesson. This important part of assessment is sometimes called moment-by-moment assessment or formative assessment. Formal assessment, also called summative assessment, is conducted as end-of-chapter tests, benchmark tests, or standardized tests/achievement tests. Both types of assessments are necessary to evaluate children's learning of mathematics. Teachers must be careful to select the appropriate assessment method that authentically measures what it is supposed to measure (validity of measurement).

10.1 Informal Assessment

The tremendous informal knowledge of mathematics children bring to school has been frequently disregarded when assessing children's mathematics knowledge and skills. Many times, children are assessed using formal mathematics assessment tools. For example, children are often asked to use the formal names of shapes (e.g., a square) when their geometry knowledge and skills are assessed. If they are unable to tell the formal names of the shapes, they do not get credit for those questions (Lee, 2014). Being able to identify the shapes using informal terms (e.g., "this looks like SpongeBob") is not taken into consideration in assessment. Informal assessment helps teachers authentically see where children are in terms of mathematics knowledge and skills. Though a child may call a square "SpongeBob," the teacher can see that the child understands the attributes of a square in an informal manner. Informal assessment often helps teachers have a better understanding of what children actually know, and this understanding can greatly supplement summative assessments.

Informal assessment will enable you to continuously improve instruction and student learning. It is more critical to assess children's mathematics learning in the early ages because their abilities to show what they know and what they are able to do are limited. Very young children have difficulty with pencil and paper quizzes or tests. If the teacher only uses pencil and paper tests, the assessment results would mislead the process of making instructional decisions as well as be inappropriate measures of children's learning. For this reason, teachers need to be careful to authentically assess children's learning considering their limited ability to express their understanding and skills.

In early and elementary levels, observation is an efficient and essential method to assess children's mathematics learning. Anecdotal notes or daily observational notes are good documentation to assess children's behaviors associated with doing mathematics. Evaluating work samples is another important way to assess what children know and are able to do. Because every process of doing mathematics counts toward getting a correct answer, evaluating the process (*how* children solve mathematics problems) is significant.

In addition to assessing children's cognitive domains associated with understanding of mathematics and skills, you should also assess children's affective domains, including their attitudes, beliefs, and disposition toward mathematics. Children's affective domains toward mathematics are likely to impact their learning mathematics, so the first step of teaching mathematics is to assess children's affective domains toward mathematics. Teaching children mathematics based on their interests maximizes their learning. For example, instead of having children memorize multiplication tables, having them make up songs about multiplication helps promote learning based on their interests. In addition, when children create a song in a group, they can use mathematics communication.

During lessons, questioning strategies can be also efficiently utilized to assess what children know about the target mathematics concept. Appropriately asking different types of questions (see Bloom's Taxonomy) will promote mathematical thinking at a higher level. In particular, interviewing a child is a good method to evaluate specific skills or understanding of concepts or processes. For example, you want to assess John's rational counting skills. You can provide him concrete materials to use as counters (e.g., cubes) and tell him to count using the materials, then ask him to give you a certain number of cubes to assess whether he has rational counting skills (e.g., "Can you give me eight cubes?"). This allows you to assess whether John has an understanding of rational counting, including one-to-one correspondence and cardinality rules.

Though formative assessment provides authentic assessment data to teachers, it has some disadvantages. First, formative assessment requires the teacher's time and efforts. Formative assessment is often time- and energy-consuming because teachers must continuously evaluate children using various assessment tools, requiring a great deal of time to collect and interpret the data. Furthermore, data are often interpreted in a subjective manner, reflecting the teacher's bias. Nevertheless, formative assessment provides teachers with insightful information regarding children's mathematics learning.

10.2 Formal Assessment (Summative Assessment)

Formal assessment (summative assessment) is often used to measure children's learning as well as to evaluate accountability. Because formal assessment utilizes standardized assessments, the results provide measurable numbers for stakeholders to see how children are doing in mathematics. Formal assessments (e.g., district or state mandatory tests) are closely associated with high-stake tests, so the level of tension about test results is often high. As a result, there are several cons about the use of summative assessments, such as teaching to the test, spending a lot of time practicing drills, and the high level of stress these assessments place on students, teachers, schools, districts, and states.

TABLE 10-1 Informal Assessment vs. Formal Assessment

	Informal Assessment (Formative Assessment)	Formal Assessment (Summative Assessment)
Description	Informal assessment (formative assessment) is an ongoing, moment-by-moment assessment of children's learning progress and process.	Formal assessment (summative assessment) aims to evaluate children's learning at the end of a semester or year. These are assessments of what children have learned.
Example	Observational notes (anecdotal notes, daily logs), checklists, participation charts, teacher-made rubrics, teacher-made tests/quizzes, work samples, self-assessment, interviews, math journals, diagrams, etc.	Achievement tests (e.g., mandatory state assessments/tests), benchmark assessments, screening assessments, etc.

REFERENCES

Blanton, M. L., & Kaput, J. J. (2003). Developing elementary teachers: Algebra yes and ears. *Teaching Children Mathematics, 10*(2), 70-77.

Common Core State Standards Initiative. (n.d.) *Development process.* Retrieved February 9, 2016 from http://www.corestandards.org/about-the-standards/development-process/

Copley, J. V. (2010). *The young child and mathematics (2nd ed.).* Washington DC: NAEYC & Virginia, Reston: NCTM.

Curcio, F. R., & Schwartz, S. L. (1997). What does algebraic thinking look like and sound like with preprimary children? *Teaching Children Mathematics, 3,* 296-300. Retrieved from http://sdcounts.tie.wikispaces.net/file/view/what+does+AT+look+like+in+preprimary.pdf

Hoyles, C. (2985). What is the point of group discussion in mathematics? *Educational Studies in Mathematics, 16*(2). 205-214.

Kilpatrick, J., Swafford, J., & Findell, B. (2001). *Adding it up: Helping children learn mathematics.* Washington, DC: National Academic Press

Kinman, R. L. (2010). Communication speaks. *Teaching Children Mathematics. 17*(1), 22-30.

Kostos, K., & Shin, E. (2010). Using math journals to enhance second graders' communication of mathematical thinking. *Early Childhood Education Journal. 38*(3). 223-231.

Lee, J. (2014). Is children's informal knowledge of mathematics important?: Rethinking assessment of children's knowledge of math. *Contemporary Issues in early Childhood. 15*(3), 293-296.

Lee, J. (2015). "Oh, I just had it in my hear": Promoting mathematical communications in early childhood. *Contemporary Issues in Early Childhood, 16*(3), 284-287.

Lee, J., Collins, D., & Melton, J. (2016, In Press). What does algebra look like in early childhood? *Childhood Education.*

Lee, J., Lee, J. O., & Collins, D. (2009). Enhancing children's spatial sense using tangrams. *Childhood Education, 86*(2), 92-94.

Lee, J., Lee, J.O.,& Fox, J. (2009). Teaching strategies: Time here, time there, time everywhere/ *Childhood Education, 85*(3), 191-192.

McGarvey, L. M. (2012). What is a pattern? Criteria used by teachers and young children. *Mathematical Thinking & Learning, 14* (4), 310-337.

National Association for the Education of Young Children (NAEYC) & the National Council of Teachers of Mathematics (NCTM). (2002). *Early childhood mathematics: Promoting good beginnings.* Retrieved April 6 2011 from http://www.naeyc.org/about/positions/psmath.asp

National Council of Teachers of Mathematics. (2008). *Algebra: What, when, and for whom (A Position of National Council of Teachers of Mathematics).*Retrieved from http://www.nctm.org/uploadedFiles/About_NCTM/Position_Statements/Algebra%20final%2092908.pdf

National Council of Teachers of Mathematics. (2000). *Principles and standards for school mathematics.* Reston, VA.: NCTM.

National Governors Association Center for Best Practices & Council of Chief State School Officers (2010). *Common core state standards (math).* Washington D.C.: National Governors Association Center for Best Practices & Council of Chief State School Officers

Reys, R. E., Lindquist, M.M., Lambdin, D. V., & Smith, N. L. (2007). *Helping children learn mathematics.* Hoboken, NJ: John Wiley Sons.

Taylor-Cox, J. (2003). Algebra in the early years? *Young Children.* Retrieved October 17, 2013 from http://www.naeyc.org/files/yc/file/200301/Algebra.pdf.

Van de Walle, J.A. (2004). *Elementary and middle school mathematics:Teaching developmentally.* Boston: Pearson Education, Inc.

Whitin, P., & Within, D. J. (2002). Promoting communication in the mathematics classroom. *Teaching Children Mathematics, 9*(4), 205-211.